IMAGINATION

The Science of Your Mind's Greatest Power

JIM DAVIES

PEGASUS BOOKS
NEW YORK LONDON

IMAGINATION

Pegasus Books Ltd.
148 W. 37th Street, 13th Floor
New York, NY 10018

First Pegasus Books cloth edition November 2019

Interior design by Maria Fernandez

Library of Congress Cataloging-in-Publication Data is available.

ISBN: 978-1-64313-203-7

10 9 8 7 6 5 4 3 2 1

Printed in the United States of America
Distributed by W. W. Norton & Company

Dedicated to my parents,
James and Janet Davies

CONTENTS

A Note from the Author ix

1 Imagination: What It Is 1

2 Perception and Memory 20

3 Imagining the Future 59

4 Imagination, Feelings, and Morality 77

5 Hallucination 90

6 Dreaming 122

7 Mind-Wandering and Daydreaming 150

8 Imagination as Mental Training, Healing, and Self-Improvement 165

9 Imaginary Companions 196

10 Imagination and Technology 213

11 How Imagination Works 229

Final Words 269

Endnotes 271

Index 315

A NOTE FROM THE AUTHOR

When I set out to write this book, I wanted to write something that bright high school graduates and university students could understand, but would also be a volume that scientists would appreciate, too.

This book uses information gleaned from hundreds of studies and scholarly works. In my field, many papers are coauthored, sometimes by six or more people. To reduce excess verbiage, I mention only the first author for a study in the text. The endnotes have all of the authors' names. This does not necessarily mean that the first author did all or most of the work, though, in general, this is true. In science, the person in charge of the laboratory is usually the last author listed.

Furthermore, most of the scientists I cite here are psychologists. If I mention a name without referring to their field, assume they are a psychologist. When I cite someone from another field, such as philosophy or computer science, I will mention it.

1
Imagination: What It Is

Imagine a jar of peanut butter.

When you do this, you're creating, in your mind, something that doesn't exist—even if you're imagining the jar you actually have in your cupboard, you're creating something new. There's the actual jar of peanut butter, and then there is a separate thing in your mind: a representation of the jar, here and now, not where and when the physical thing actually is. The actual jar of peanut butter is made of plastic and peanut butter. The thing in your head, the "imagining," is some pattern of neurons firing in your brain. Even when you use your imagination to remember something that actually happened to you, you're creating a simulation of a time and place that no longer exists.

This is the essence of imagination: the creation of ideas in your head, composed from ideas, beliefs, and memories. Often, they are not simple ideas, but complex structures. The most spectacular use of imagination is in creativity, but this book isn't about creativity, which requires the generation of something new and effective in some way. Acts of imagination need not be new or useful. Imagination also has great uses in more mundane tasks we do every day, such as planning the day. When you think of what route you want to use to get home, or you go through the logistics of where to park your bike, or figure out what order you should run your errands in, you're thinking of possible realities that do not yet exist. Though we don't often call these acts "creative," they are fantasies, possible futures that don't currently exist outside your mind. Even the simple act of considering what to do next is using your imagination.

When you picture a jar of peanut butter, if you're like most people, you have an experience that is kind of like, but not exactly the same as, seeing it in real life. Likewise, when you have a song stuck in your head, or when you're having a vivid dream, it can be a profoundly sensory experience. It's called your "mind's eye," because it feels like you're actually seeing things in your head. Likewise, we have a mind's ear, nose, and tongue.

MENTAL IMAGERY

For most people, imagination in its clearest and most obvious form is mental imagery. You have the experience of "seeing" in both perception and in imagination, but in perception, the light or sound comes in from the outside world, and for imagination, the information comes from your memories.

Coming up with a precise definition of mental imagery is difficult, but coming up with a vague definition is pretty easy. Mental imagery tends to have some common characteristics: first, it is *like* experience with the senses. That is, if you create a mental picture of a boy chasing a fox, or imagine the sound of what the fox says, the experience is like a less-vivid version of actually looking at a boy or listening to a fox in the real world. Second, it happens in the absence of input from the environment that would normally cause it—it's created by your own mind.[1] Or, put more simply, the image in your head is not caused by what you're currently perceiving.

This bit is important because even when your eyes are open, your mind can rope in memories and project imagination into the scene you're seeing in the world. You might look at your living room and imagine how a red couch would look in it, in a kind of organic augmented reality. Imagining doesn't have to be in the absence of *all* perceptual imagery, just in the absence of the stimulation that would *normally* produce the experience—in this case, an actual red couch.

In these examples, we generate imaginings with an act of will, but sometimes imagery comes automatically. For example, many people get spontaneous images when they're reading novels. I've described disturbing things to people to have them tell me, sarcastically, "Thanks for that image." People who experience trauma sometimes imagine the event, again and again. Not only do they not decide to reimagine the trauma, they can't stop.

But can we imagine without imagery? It's easy to think of visual imagination as being nothing more than mental imagery, but we have what we might call "conceptual imagination" as well that doesn't really have much to do with the senses. I'll give a few examples.[2]

Imagine a triangle. Now add one side to make a square. Now add so many sides that there are 2,001 of them. The picture of it in your mind's eye is wrong—either the polygon has far fewer than 2,001 sides, or it looks just like a circle, because the details are too fine to make out with the resolution of your mental imagery.[3] Just like a computer screen or a photograph, your visual mental imagery has a limited resolution.

So how is imagining a 2,001-sided polygon different from imagining a circle? Because you *know* that it's a polygon, not a circle. Now, remove one side of the polygon, so that it only has 2,000 sides. It doesn't look any different in the image! Both a 2,000-sided polygon and a 2,001-sided polygon will look just like a circle in your mental imagery. The difference is only in your belief about the polygon. Both of them look like circles, but you *know* that they have a different number of sides. These beliefs are part of your imagination, too, even if they don't particularly look like anything. The difference is conceptual, not visual. This is one example of how you can have a nonsensory imagining.

There are lots of states of being that don't particularly look, sound, or smell like anything at all: owning a bike, thinking that *Kiki's Delivery Service* is a great movie, having $49,000 in your bank account, having a goal to chew more gum, wanting to eat a spoonful of peanut butter, being part of the in-group at work, and so on. Often, our imaginings are a combination of mental imagery (sensory imagining) and conceptual imagining.

And this ability is extraordinarily powerful.

Our ability to imagine things is a surprisingly important ingredient for the special sauce we have that makes us the only species on Earth that have things like money, the arts, cities, moon landings, and Laurie Anderson's *Big Science* album.

Let's take social organization as an example. Many of our relationships with other people are one-on-one relationships. You know your lovers, your friends, your parents, your rivals, the other people at your workplace, and so on. Many other animals also have these kinds of relationships. Even bats remember other individual bats in terms of who did and who didn't help out when food was scarce, and respond accordingly, punishing the jerk bats.[4]

But we humans have lots of relationships that go beyond our *personal* histories with other people. We can feel a kinship with (or a hatred for) people we've never even met. We might feel a nationalistic affiliation with our fellow citizens. We might want to help out other people who belong to our own religion, political party, or are also Beastie Boys fans—or hurt those who we believe hold different values. Other animals can't do this. Chimpanzees can't have groups of more than about fifty individuals. Any more than that and the group splits. Their brains don't have the carrying capacity to keep track of that many one-on-one relationships. Nonhuman animals can't know who to trust without having some significant interaction with specific individuals. But with human imagination, we can construct these abstract concepts of social groups that include people we've never met. When we perceive someone else to be in some social group or other, we know how to treat them.[5]

The very idea of a social group that includes people you've never met requires using conceptual imagination. Social groups have no physical existence, or, more accurately, they exist only because lots of people believe they do. Without this imagination, we could not have money, or countries, or religions.[6]

We can see other examples of conceptual imagination in our dreams. You might dream of someone who looks like a nun with a nicotine habit that you *know* is your mother. From a sensory perspective, it's a nun. You knowing that it's your mother is using a nonsensory imagination.

So we can imagine without using imagery, but it's a little bit harder to get our minds wrapped around this. If you're like most people, when I ask you how many windows were in your childhood home, you will do a mental walkthrough of the house, counting the windows as you go. But if I ask you whether the roof of a house is above the door of a house, you can probably answer that without making a mental picture. You *might* make a picture, but you probably don't need to. Why do you use mental imagery for one and not the other?

The fact that roofs are above the door of a house is just that, a fact. In psychology, it's what we call a "semantic declarative memory." It's sentence-like, but it's not about any particular event. Now, when I say sentence-*like*, I don't mean that it's stored in your brain as a sentence in the language you speak. But it's *like* a sentence in that it has symbols that are arranged in a meaningful structure. Think of a pair of shoes you own. You know you own them, but the fact that you do doesn't look or sound like anything in particular. It's just a fact about the world. You can picture your friend borrowing and wearing your shoes, and that image doesn't make the shoes theirs. Knowing someone is jealous, or married, or likes the taste of peanut butter, also doesn't look like anything either,[7] but you can imagine these things easily.

Just about all psychologists believe that *all* of our long-term memories of things are connected symbols. This is true even for your visual memory. It's made up of symbols for objects and shapes, textures, distances, and so on, and the relationships between these objects is encoded in these sentence-like entities. In cognitive science, we call them "propositions." But it certainly doesn't feel that way. It feels like we have pictures in our memory that we can recall pretty well. So why would psychologists think that our long-term memories don't have pictures?

Think of the last large dinner party you attended, and picture it in your head. Doesn't it feel kind of like you had that picture somewhere in your memory, and you simply brought it to mind? That the picture was in your head, fully formed, and you are just directing your attention to it? It kind of feels like pulling a photograph out of a drawer and looking at it. But as much as it feels like this, it's almost certainly not what happens—that would be like having a photographic memory, which is exceedingly rare, if it exists at all. If someone you know thinks they have a photographic memory, find a book they've read recently, open to a random page, and ask them to tell you the last five words on that page. They won't be able to do it.[8]

Suppose you reminisce with a friend, and your friend reminds you that someone you'd forgotten about had also been at the dinner party, your forgettable friend Erika. That is, the image you saw in your head was missing

Erika. If your long-term memory of that scene were really like a picture, there wouldn't have been a person missing from the mental image.

A true photographic memory might make some mistakes, but the mistakes would be of a different kind than those we *actually* make: Think about what was in that image where Erika *should have been*. It's not like a blurry or pixelated part of a photo, or a part ripped out, or a big black space. Those are the kinds of errors you get if that person were missing from a photograph or a digital image. Rather, your mind made (what you thought was) an apparently complete, coherent picture of the dinner scene. There wasn't someone obviously missing from it. You'd have noticed something like that. Instead, you just omitted Erika, and the picture looked fine and complete to you in your head. Which is really kind of amazing. You reconstructed a picture, or something like a picture, from nonpicture memories. You just failed to recall Erika, so she didn't get cast in your inner picture.

So what's going on here? To the best of our knowledge, you have a long-term memory of the dinner party that is interpreted, made of symbols connected in propositions. When you are called upon to engage in memory recall, your mind can *construct* a mental image based on the information in long-term memory, the fact-like information about who was there, where they were seated, what was for dinner, and so on. It's a little like a description you might give to an illustrator. Because individual facts can be forgotten, the resulting picture might be experienced as complete, yet still have inaccuracies.

Then, when somebody reminds you about Erika, you just slip her seamlessly into an already pretty good picture. It's like doing a great Photoshop edit in your mind. It feels so natural and happens so effortlessly, we hardly notice that it happens this way. But this is important, because it means that *even what feels like simple memory retrievals are, in some sense, acts of imagination*.

When you update the facts, a new picture emerges. And you might remember this new picture as what happened—but you never store the image itself, just another description of it.

Of course, we don't *have* to use mental imagery. Sometimes you can rattle off a list of the people at the party without making a picture of it. You are still engaging in imagination. The mental imagery part comes afterward and is, in many cases, optional.

When I talk about imagination, I'm not just talking about imagery—you can *imagine* things with or without *imaging* them.

I should say here that although scientists agree that people can *experience* mental imagery, there is some disagreement about what's actually happening in your mind when this happens.

PSYCHOLOGICAL DISTANCE

Although most of us can generate (and be conscious of) pictures in our heads, we don't always do it. Suppose you're planning errands with someone you live with and discussing which of you will go buy that jar of peanut butter. Many people can engage in planning like this without actually picturing anything in their heads. But if they choose to, they can. Sometimes we can turn a conceptual imagining into a mental image with an act of will. You can generically imagine having a jar of peanut butter, but if you create a mental image of it, you need to make a picture with a specific place, with a certain point of view—something that is unnecessary when you're just hypothetically imagining "I have peanut butter."

So when do we engage in mental imagery and when do we just stop at conceptual imagination? It depends, in part, on what is being imagined. Imagining some things makes us more likely to create mental imagery than others. For example, we tend to think more in pictures for things that seem physically close to us. If you talk about a (hypothetical) birthday party that's nearby, people will have more vivid imagery than if you're talking about the same birthday party that's really distant.

But this closeness can also be metaphorical. There also seems to be an association between actual distance from you and what we might call "psychological distance." A good example of something that is psychologically distant is an event far in the future. Elinor Amit has a theory that we think more in pictures for things that are psychologically close and think in terms of words and propositions for things that are more psychologically distant.

To test this, she used a version of the Ponzo illusion, which involves two diagonal lines that converge near the top. Two equal-length horizontal lines will appear to be of different lengths depending on how high they are placed on the image. Think of a photo of train tracks going into the distance: even though each railroad tie has a different length on the page or screen, they look to be the same length because the rails give the impression of lines converging in the distance. Your mind adjusts for the foreshortening that you expect from distant objects.

In her experiment, she placed words or pictures either high or low on the image of the two converging lines. Then she measured how quickly people could recognize them. Things higher in the image look farther away, and things near the bottom of the image look closer. People were relatively faster at recognizing words when they appeared far away, and pictures when they appeared close.[9]

In other studies, she asked people what they would prefer for bathroom labels: iconic pictures of men and women (the kind we see on restrooms all over) or the words "men" and "women." If the signs were to be put up next week, people showed a preference for the pictures. If the signs were to be put up a year from now, they preferred the words. She also found that people were more likely to use pictures (versus words) with people perceived to be socially close to them, and words for those further away—you're closer to your mother than to the CEO of Disney.

These distance effects work for physical distance, distance in time, and even social distance. People are even more likely to follow sent instructions when the medium and distance are congruent. That is, if you're communicating instructions to someone about how to deal with something close to them, they are more likely to follow those instructions if you present them in pictures. If the instructions regard something far from them, you're better off using words. Effective communication depends on whether your use of words or pictures corresponds to how distant the subject matter feels to the person you're communicating with.[10]

I find these effects pretty remarkable. It makes sense that our minds would naturally think in terms of space: we, as well as most of our animal ancestors, need to navigate space to survive, so it makes sense that we'd be pretty good at it. But it seems like we use the concept of distance in space to help us understand more abstract ideas like time and social connection. It's a very deep metaphor. This stuff doesn't only work for distance. It turns out that human minds work with metaphors all the time.

METAPHORS AND THE MIND

Let's talk about how we think about time. Suppose your coworker tells you that the meeting on Wednesday has been moved forward by two days. When do you think the new meeting is going to be held? On Monday or on Friday?

The answer might seem obvious to you, or you might not be too sure. People differ on whether they think the new meeting is going to be on Monday or Friday. What's interesting is that the one they pick depends on how they imagine time, and what metaphor they use.

Some people think of time as moving around them. It's as though they are standing still in a fast-moving river, facing upstream, and imagine time flowing around their bodies. So the idea of an event in time being moved "forward," means, to them, closer to the present, coming up sooner, and earlier in the timeline. These people tend to think the new meeting is happening on Monday.

Others view time as static, like the still water in a pool, and they are the ones wading through it. To them, "forward" is the direction they're moving in, placing the new meeting on Friday. So be careful when you tell people you're moving an event forward or backward in time. The different metaphors for time, which they don't even realize they have, will make them interpret the new date differently.

But it's not that each person has only one way to think about time. We all can imagine time with either metaphor. We understand what it means when we hear things like "Winter is coming," which suggests that time itself is moving, and we also understand things like "when we get to the end of the year, we should have money saved up," which suggests that we're the ones moving through time.

What we're doing can affect which metaphor we choose. Lera Boroditsky asked people in lunch lines whether the Wednesday meeting pushed forward two days was on Monday or Friday, and their answers seemed to depend on where they were in the lunch line! People at the front were more likely to say "Friday." Those at the back of the lunch line were more likely to say "Monday."

Boroditsky's explanation for this is that people at the front of the line feel like they're moving. We've all felt this. At the end of the line you feel like you're not getting anywhere, and as you get closer to the front, the movement is steadier. So it seems that the ones at the front had their minds primed to think of the self-moving metaphor of time, while the people at the end of the line, who felt like they were barely moving at all, felt like they were just waiting for time to pass.[11]

Notice that for both of these metaphors, people are thinking about time in terms of space. This is just one example of how we think about abstract things like time in terms of more concrete things, such as handling physical objects. We do this a lot. More abstract things tend to be understood in terms of more concrete things. Our imagination affects our understanding of these concepts.

So why do our minds think in terms of metaphor so much?

To understand why, I need to tell you a bit about brain evolution. If we look at lizards, and other creatures that have simpler brains than humans, we can see that they have brain structures similar to humans. They have limbic systems for running basic fight and flight mechanisms, perception, body control, and so on. Lizards don't have language, so they don't have the brain areas that are specific to language.

But humans diverged from lizards in the evolutionary tree a very long time ago, and we've both continued evolving since. So why would we still have recognizably similar brain structures?

If you want a new boat, you can tear down the old one and build a new one that's completely different. But evolution can't completely overhaul the brain plan of a species in the same way. As a brain changes over time, those creatures still have to be able to stay alive. To continue the analogy with the boat, it's as though you're in the boat out on the ocean and you have to change it, but you can't make changes that would make it sink *at any moment*. What this means, for boat building, as well as for brain evolution, is that a lot of the brain parts that are necessary for survival *can't* be changed all that much. Evolution works incrementally, and you can't tear down your ability, say, to digest protein and build a new one, because you have to digest protein. All the individuals in the population would die and wouldn't be able to reproduce.

The situation for evolution is even harder than for a boat. You might be able to build something on deck and not use it until it's finished, such as a new sail. You put the sail up only when it's done and probably going to work. But evolution can't look ahead like that. A system can't evolve unless it's useful *every step of the way*. So, in this analogy, you would have to subtly make changes to the sail, making it incrementally better with every single change.

In evolution (of anything, not just brains), just about nothing will stick around unless it's useful. That's why wings didn't start out as wings, our ear bones didn't start out as things we hear with, and so on.

Our brains can't stop using the functions they need to keep us alive and reproducing, and evolution can't create new functions unless they are useful every step of the way. So instead of rebuilding the brain, evolution ends up building additions to what's already there. There is no "closed for renovation" in living species.

The brain areas at the top of our brain stem (the bottom of the brain) are the oldest. Things like sex, escaping predators, perception—evolution can't really mess with these systems, so it builds new ones and tacks them on like extensions built onto a house. For us, these additions happened on top and in the front. I'm speaking as though evolution has a plan in mind, but, of course, it does not. It's easier to explain evolution if we treat it as a designer with constraints, but eventually you have to let go of that scaffolding, because evolution can't plan anything.

These new structures are added onto the old ones and, indeed, *use* those old ones without changing them much. And when you have several new structures relying on the behavior of a single old one, changing the old one becomes impossible. Too many other systems are relying on it working exactly the way it does. If eight systems are using one more–primitive system, then all eight are expecting that old system to work with particular inputs and outputs, and changing that old system would cause a catastrophic system failure.

We can think about it in terms of a company using an outdated computer or software. A company might get a database system, and over the years, more and more parts of the business are created that need to access and use it. These new systems are designed to efficiently make use of the way this database software works. But when new, better database software becomes available, the company sticks with the old one. Why? Because so many other systems in the company need it to work the way it always has. To upgrade the database software, all of the other systems would also need to be changed before the company could run properly.

It's the same with evolution. The perceptual systems are like that old database software. Newer parts of the brain, those that evolved fairly recently in evolution, which we share with fewer animals, use the outputs of the visual system in a very rigid way.

Should the old system change, the newer systems would need to change at the same time. A company can, at some expense, upgrade *all* the software at once, changing everything and creating a whole new information processing system, but this is something that is just about impossible in evolution. Note that the opposite is not true—one of the new systems can change without the old system changing, because the old system does not depend on the inputs and outputs of the new systems. As a result, newer brain systems can evolve more rapidly than older ones.

The old brain is a complex hodgepodge of special-purpose mental machines. There's an area for recognizing faces, for detecting color, for feeling basic emotions, and so on. It's efficient, fast, and rigid.

The new brain is much slower, more conscious, and does things systematically. It is specialized in deliberate action. It behaves more like a general-purpose learner. When you are doing long division on paper, you are heavily taxing your new brain. The new brain is more frontal, where old brain structures tend to be at the top of the brain stem, in the middle of the brain, and closer to the back.

Another distinction between these old and new systems is that the new brain tends to work on things one at a time, step by step, and the old brain works in parallel. As a result of this, the old brain is much better at taking in a great deal of information at once and responding to it. The new brain is just too slow for so many things at once. For some tasks, it's better to rely on new brain processing, and for others, it's good to use old brain processing.

So "older" brain areas are ones we share with other animals. They're not exactly the same, of course, but they are similar enough that we can learn a hell of a lot about the human brain by studying, say, cat brains. The binary classification of "old" and "new" is a bit crude. It's more of a continuum, starting in the middle of the brain, at the top of the brain stem, and curling to the back and over the top to the forehead, going from older to newer. But for the sake of explanation, referring to processes as old and new can be clear and instructive.

Neuroscientist Michael Anderson tested this new/old brain idea by looking at large-scale brain activity. It turns out that when you are using old parts of the brain (emotion, motor control, vision) only those parts are excited. (The whole brain is active all the time, so we're talking about relative activation here.) Because those old brain areas evolved before the new ones appeared, they don't need to use the new ones to work. But the new brain areas are just the opposite—they use the brain areas that came before them, kind of like using tools. So when you look at language use, which evolved very recently, the whole brain lights up, but for simple vision tasks, the activation is restricted to older areas in the back of the head.[12]

When people talk about your "lizard brain," it's more literally true than you might think. So that's why we can talk about the old brain and the new brain. The old brain structures are like old buildings on a college campus—still kept up and modified, but their basic structure was determined long ago, and can't change much.[13] The old brain is also more important. If you're going to get hit in the head, you're better off getting hit in the front than in your lizard brain. It's hard to live with a reptile disfunction.

The visual system, and thus, a good part of the visual mental imagery system, is very old, because our ancient ancestors needed to be able to see more than they needed to be able to concoct long-term plans. As such, it works mostly with other old systems like emotion, and things that deal with the world in front of us, as opposed to hypothetical notions. In contrast with conceptual imaginings, visual images are concrete, visceral, and emotional. As we might expect, people who make more vivid mental images also experienced stronger emotions, as found by my colleague Eve-Marie Blouin-Hudon.[14] So perhaps it should not come as a surprise that visual representations are more emotional and related to more concrete, close things. Vision is part of an old, primitive part of the brain, and visual imagery associates more closely with other older parts of the brain.

Now that we have a basic picture of the brain, we can return to metaphor. Why might the brain think of time in terms of space? In general, non-human animals think about simpler, more concrete stuff like jars, drinking, and geese. As humans evolved to think about more abstract concepts, we tended to use concepts we already had—that is, more concrete ideas, such as space and our navigation through it. That's why we think about love as a journey, and up being good, and darkness being bad. We think about abstract concepts in terms of concrete ones, and that helps us understand why we imagine time the way we do.

When we think of sensory imagination, we tend to think of pictures in the head. Indeed, of all of the senses, sight is the most important for human beings, and most of the discussion in this book will be about visual imagination. But technically, we can image in any of the sensory systems we have.[15] Imagery in our minds can be multimodal.

We are all familiar with auditory imagination. We get songs stuck in our heads, and about one quarter of conscious experience is "inner speech," or talking to yourself in your own head.[16] Rather than pixels, sound is imaged as nonsymbolic bits of sound. Similarly, we have imagination for smell, touch, and taste, though most people cannot imagine smells with any vividness. I know what peanut butter smells like, but when I try to imagine that vividly, I can't do it. Your mind's eye is sharper than your mind's nose.[17]

We also have what's called "motor imagery." Your motor system controls your muscles, and motor imagery is when you imagine doing something with your body without your body actually moving. You might dream of running, or mentally practice a dance routine, or vividly imagine going out and buying several copies of this book to give as gifts. In these cases, your motor system is going through the same processes as it would if you were actually doing these things, but the signal gets stopped before it reaches your muscles. We use motor imagination to anticipate other people's actions,[18] and some have argued that when we move our bodies, we are often playing out an action sequence already planned out in the imagination.[19]

Most people immediately know what I'm talking about when I mention pictures in the mind or other kinds of imagery. But if you're someone who doesn't, you're not alone. People vary in the amount of visual mental imagery they experience, and the detail and vividness of those images.[20] Some people can have imagery so vivid that they sometimes have trouble telling their imaginings from reality. Neuroscientist Adam Zeman calls these people "hyperphantasics."[21]

Although some scientists claim that imagination is central to consciousness,[22] many people (including two from my imagination laboratory, ironically) report that they have no conscious experience of imagery at all, even in dreams.

As mentioned above, most people, when asked to count the number of windows in their childhood home, do a virtual walk-through in their mind's eye, counting as they go.[23] "Imagers" report that they have a vision-like experience in their minds, and that they can read information off of this image similarly to how they do it with things they're seeing in the world.[24] Most people are vivid imagers, and only about 3 percent of people have very low imagery ability. Some people start out as imagers, but, usually due to brain damage, no longer experience these images. One woman reported that, after a car accident, she had trouble understanding speech because she could not picture what was being said. But after six months she had learned to associate words with auditory imagery and regained her full comprehension abilities.[25]

APHANTASIA AND HYPERPHANTASIA

Although we've known about differences in imagery vividness between people for a very long time, the phenomenon of having no imagery experience at all has only recently been appreciated, so we don't have lots of data on it yet. Adam Zeman got some media attention for his research on what he called "aphantasia," and as a result, twenty-one people contacted him, telling him that they also had no imagery experiences. Of these twenty-one people, nineteen were male. This gender difference might be a function of the readership of the magazine where the original article appeared, *Discover*, which skews male. For ten of them, *all* senses were affected. That is, they could not generate imagery for vision, sound, smell, or anything else. Nine had "substantial" reductions, and ten could generate absolutely no imagery at all, though most had occasional involuntary images, and seventeen had imagery during dreams. When asked how they accomplished tasks that more people used imagery for, such as reporting how many windows were in one's house, they reported using knowledge and memory, rather than visual information.[26]

Learning about this feels revelatory for many people, because we tend to assume that everybody else has the same experiences we do—and we assume that experiences we don't have, others don't have either. Aphantasics assume that nobody has imagery, and people with really vivid imagery have a hard time picturing anybody being different. This even extends to scientists, who are more likely to believe in scientific theories that match their own experience with imagery.[27]

An aphantasic from my lab said ". . . not only is it possible to imagine something without images, but that's about all we can do. I don't consider myself as a person leading an impoverished internal life, nor do I think of myself as not creative or unimaginative. I simply don't have the pictures." She also said that she has no auditory imagery either—for example, she can't "hear" a song in her head when she imagines it. You know how sometimes you get a song stuck in your head? Never happens to her.

A friend of mine, the philosopher Jeanette Bicknell, also appears to have aphantasia. She read an article about it and told me that "on the question where you were asked to imagine a thunderstorm, I could 'hear' thunder and 'felt' shivery from the rain, but I didn't actually 'see' much. And I imagine music so vividly that it is like really hearing it. Ian [her husband] told me that when he imagines a visual scene, it is like he is replaying a video tape that he took previously. I've never experienced anything like that. And (like the guy in the article) when I imagine a person, I imagine facts about them. . . . Like another person quoted in another article, I also remember directions as a set of instructions. (Actually didn't know there was another way to understand them.) And I used to get frustrated by the visualization exercises at the end of yoga classes, so I started tuning them out."

Jeanette's experience shows that aphantasia is not necessarily cross-sensory. She has vivid auditory imagery (in fact, her specialization is in philosophy of music!) and tactile imagery. She is *only* lacking in the visual. So aphantasia isn't all or nothing.

One task that seems to *require* imagery is to ask someone what shape is made in the negative space of a capital letter A. Most people use visual imagery to do this, but aphantasics can do it, too, they just do it with motor imagery, imagining drawing the letter A with their hand and noting that it feels like drawing a triangle. But because imagery is completely an internal experience, and people can often use more than one kind of imagery to do a task, this area is very difficult to study. For example, we might want to study how someone without visual imagery does on a task. But they end up doing well on the task because they figure out a way to use some other kind of imagery to do it.[28]

Imagery scholar Bill Faw writes that when he tries to image things, it doesn't feel quite like conceptual imagination, but more like "subliminal imaging." He writes: "When I am trying to 'picture' the face of my wife (now of 48 years!), for example, I try to remember seeing it—but don't 'see' it. It is almost as if I try to draw her profile, nose, and mouth on water—sort of outlining it but leaving no visual trace . . . "[29]

But most people have *some* amount of imagery experience, and they differ in how vivid it tends to be. That is, the capacity for the vividness with which you imagine things tends to be stable over time,[30] even though one person's imagery might be more or less vivid for one imagining versus another—depending, in part, on effort, the subject matter, and whether there is distracting information from the environment. For example, it's harder to have a vivid visual image if you're watching a movie at the same time. I ride my bike to work, and while I do, I sometimes sing albums to myself. This isn't difficult, in part, because the parts of the brain used for singing and verbal communication aren't really important for cycling, which instead requires great visual and motor attention. One of the albums I sing is Paul Simon's *Graceland*, and I have a visual memory cue to help me remember the track order and lyrics. But I can't really vividly visualize these cues very well, because when cycling your eyes are highly stimulated with a constantly-changing environment. Because visual perception and visual mental imagery use many of the same brain areas, I get interference. This is why it's often easier to image something with your eyes closed, or when you look up at a featureless ceiling.

On the other end of the spectrum from aphantasics are the superimagers, the hyperphantasics, whose mental images are experienced as extremely detailed, vivid, or accurate. Good imagery is typically thought of as being vivid, or subject to excellent control by the imager. Powers of vividness and control usually go together, though through injury some people still have imagery, but can no longer control them.[31]

So, it's clear that people differ in their inner experience. Some people have very vivid sensory-like experiences in their imagination, and others have none at all. Most of us fall in between.

When we talk about imagination, we're talking about generating something in the mind. At its most basic, every memory recall is imagination, because memories are reconstructed every time they are retrieved. But better examples of imagination are when one generates a hypothetical situation—maybe a plan for the future, or a fantasy. These imaginings start out as fact-like ideas, and then they *might* be turned into sensory-like experiences we call "images." When I talk about imagination, I'm talking about all of these. When I talk about imagery, I'm referring only to the creation of what is experienced as sensory-like imaginings.

It looks like what happens is that when you imagine something, you are drawing on and reusing memories to form something new to think about. Because imagination is so dependent on the memory and perception systems, let's get into them a bit.

2
Perception and Memory

Your imagination system has a lot to do with your perceptual and memory systems, so I want to spend some time talking about how they work.

As you go about your day, your eyes are open, and you effortlessly understand most of the information going into them. It feels like the simplest thing in the world. But, like the song "MMMBop" by Hanson, the apparent simplicity is an illusion.[1] The human vision system is really complex. It just doesn't feel complicated because our conscious minds are spared just about everything except the final product. It's kind of like buying a jar of peanut butter. It feels simple, but that's because you don't have to think about growing peanuts, harvesting them, shipping them to a processing plant, turning them into peanut butter, the manufacture of a plastic container, the printing of the label, the shipping, and so on. Your interface only needs to be a store and some money. Similarly, your conscious mind enjoys a rich interpretation of what's in front of you without having to concern itself with the details of how it was generated. But there's a lot going on under the hood.

Light enters the eye and gets focused onto the retina, a screen of cells at the back of your eye. You're probably used to hearing about neurons in your brain, but you have neurons in your eyes, too. They are sensitive to light—that is, when light hits them, they fire more frequently, or, as we often say, they get more "active." When we say a neuron "fires," what we mean is that it's sending a signal on, usually to another neuron. Neurons communicating through firing accounts for the lion's share of how the brain works, at a cellular level.

But what they're communicating is *information*, which has no fixed biological reality. Information is abstract, which can make it a little harder to understand than, say, the processes of your circulatory system. We're not going to get anywhere talking about this neuron, firing or not firing, and causing or preventing firing in the next neuron. It's like trying to describe how a word processor works by talking about circuits in the computer. So we talk about information. If you see the edge of an object, it's the information about that edge that gets passed around the brain. It's done by neurons, of course, but it's more comprehensible to talk about the information rather than the biology of it.[2]

And even this early in the vision system, retinal neurons are processing some of the information, turning it into something useful for the later neurons.[3] From there, the information goes through several brain areas, each of which does some kind of processing to it, extracting important, new information from it. The next bit is a little technical. The brain is intimidatingly complex, even more complex than "MMMBop," but I have to describe at least some of the complexity to preserve this book's scientific integrity. You don't have to understand how the brain works to get a lot out of this book, so just skim it over this next part if it gets to be too much brain for your brain.

One of the first places image information goes is called "visual area one," or V1, in the back of the brain.[4] The information that was on the retina is more or less reproduced here. We know this because of an experiment that was done on monkeys, who have visual systems a lot like ours. Neuroscientist Roger Tootell trained a monkey to look at a pattern of blinking lights, and injected the monkey with a radioactive sugar. The idea is that neurons that are more active will absorb more sugar, because neurons consume sugar and oxygen as fuel. Then they sacrificed the monkey (that's science-talk for "killed it for the good of science") and looked at its brain, specifically area V1. Then they read where the radioactivity was most active. In effect, this shows us a picture of which neurons were doing a lot of firing. Looking at the picture, you can easily see the pattern the monkey was looking at (a grid pattern), actually visible on the slice of brain. What this means is that the pattern of activation caused by light on the retina is re-represented more or less intact in V1 for further processing.[5] Stephen Kosslyn calls this area the "visual buffer" ("buffer" is a computer-science term for a temporary memory).

Now your eyes are always moving. The big moves you can feel, but the little moves, called "saccades," are unconscious and frequent. This means that your retinas are getting a completely new picture about forty times every second, and V1 is, too. So the images re-represented in the visual buffer quickly fade—the reason the monkey experiment worked was because the sugar's radioactivity takes a long time to decay. But the increases in activation last a very short time. If they lasted very long, we'd see a smear every time we moved our eyes, like trying to fill a whiteboard with new diagrams before erasing the old ones.

What does this all have to do with imagination? Well, V1 is also the site of visual mental imagery. That is, when we image something visually, it shows up in V1 as a pattern of activation that resembles what it would look like on the retina. But it lasts a very short time. If you want to keep a mental image in mind, it needs to be frequently refreshed from a longer-term memory, just like a TV screen needs to be constantly refreshed by a signal from elsewhere to keep a picture.

Back to vision. V1 is just the beginning of visual information's long safari through your brain's jungle of neurons. Visual perception re-represents information in one brain state after another, and the further along it goes, the less the pattern resembles the original picture on the retina. It gets more and more abstract. Broadly speaking, from V1, the information takes two paths—one for visual processing and the other for spatial. Visual processing (done in the inferior temporal lobe, near the bottom of the brain) deals with shape, depth, color, intensity, and object recognition. Spatial processing (done in the posterior parietal lobe, near the top of the brain) deals with orientation, size, and where things are in space—either objects in space, or where the parts of a single object are in relation to each other. The visual path is casually referred to as the "what" pathway, and the spatial path the "where" pathway.

People with brain damage to one of these paths have trouble with those corresponding parts of perception. For example, if someone has a problem with their temporal lobe, they might know that some object is in front of them (because the spatial system is working) but have no idea *what* that object is! Conversely, someone with a problem in their parietal lobe might know what's in front of them but have no idea where it is.

Medical scientist David N. Levine describes a patient with damage to his "where" pathway. He could name the colors of large objects, but could not track them with his eyes. He could name objects and colors in front of him, but not reach for them, nor describe where they were. If presented with two things, he could not tell which one was closer, or above, or farther left.[6]

In case that didn't startle you, let me say it again: there are people with neurological problems who can know what object they are looking at, but have no idea where it is. For people with intact visual perception, this is hard to imagine, because our visual experience feels so seamless—we really don't know what it's like to look at an object and know what it is without also knowing where it is, and it certainly doesn't *feel* like two completely different parts of the brain are dealing with "what" versus "where" information.

But you can get a hint of knowing where something is without knowing what it is. The neurons that take care of the outskirts of your visual field specialize in motion detection, and not so much in high-resolution details. Try this: look straight forward at the wall or whatever's right in front of you. Put your arm straight out to your right side, back so far that you can't see your hand. Then, wiggling your fingers, slowly bring your arm forward so that your wiggling fingers gradually come closer to the edge of your vision. Do it slowly, and don't move your eyes. At some point, you will get a sense that motion is there, but you won't be able to really see your fingers enough to, say, know how many you are wiggling. You're seeing motion, seeing that something is there, without seeing what it is. People with certain disorders have experiences like this, but for their whole visual field—they have vague notions of where things are, but not what they are.

Information goes though these processing stages, and the further in it goes, the more complex perceptions the mind pulls out of it. Eventually, the mind knows what objects are there and where they are. It tries to find matches between the information coming in and stored memories of what things look like. When there's a good match, the mind says, in effect, "Aha! There's a jar of peanut butter!" It also can detect events, such as doors opening, and it is combined with how we understand people, such as when we perceive someone as trying to open a tight jar.

This process, as I've described it, is "bottom-up:" how the mind extracts information from a raw stream of data from the world. Our perception feels like it's driven completely by what's out there in the world. But actually, what we think is out there is a complex combination of what information is coming in from the world and our understanding and expectations of what the world is like. There are also "top-down" processes. It's easier to see what you expect to see, and the mind is constantly guessing at what will come next, and priming early visual areas to make seeing those things easier to do.

For example, when you open a book, you expect to see pages of text. Imagine that you opened a book, but instead of seeing pages of text, you saw an actual jar of peanut butter. It would take you a little longer to even register what it was, and you would react with surprise. Surprise, for sensory information, is a violation of the expectation system in your head.

While you're listening to someone speak, if a loud noise happens in the middle of a sentence, chances are you'll still understand what they're saying. Suppose a firecracker goes off with a "bang": "I thought the peanut butter we had on our toast this morn-BANG was really good." You would experience hearing the whole word "morning" even though the end of it was obscured by the firecracker. In vision, too, your expectations will "fill in" information missing from the world to help you make sense of it. When you see someone behind a railing, your mind understands that it's a whole person there, and it's not that the person doesn't exist where the railing obscures your view of them. These are "top-down" effects.[7] When we notice the colors of things, we later remember those colors being more prototypical than what we actually saw. That is, we might remember an unusual shade of yellow as being closer to a school-bus yellow.[8]

Similar processes happen with all of our senses, and we can see examples in our perceptions of music. In the Western tradition, we have a certain set of notes that we use most of the time—other cultures use different sets. When Westerners hear a note that's just slightly off pitch, our minds change it to something that's on pitch. That is, even though what we might hear is slightly flat of a middle C, we will actually experience a perfect middle C.[9]

The same thing happens with words. For example, if I told you my brother-in-law played the "kitar," you'd probably hear the "k" as a "g" sound, because you're expecting the sound "guitar" at the end of that sentence. In a sense, you misheard me, so that's getting it wrong. Right?

Well, maybe. Another way to look at it is that your mind is doing a kind of auto-correct. I probably meant to say "guitar" and just misspoke, so you not even noticing that I *actually* said "kitar" doesn't really matter—it helps you, in fact, so we can both get on with the conversation without you having to ask what the hell a "kitar" is.

So what we perceive in the world is a combination of what's coming in through the eyes and what we expect to see with them. One way to think of these top-down patterns that aid in perception is to think of them as symbols.

THE IMPORTANCE OF SYMBOLS

As we age and experience more and more of the world, we create ways to classify things, and to label them with symbols, which get used for expectation. If you look at some travel photos, it's easy to pick out people, beaches, umbrellas, ocelots, and other things. When we recognize a particular patch of light as a jar of peanut butter, we store it in memory as a symbol—an instance of a jar or peanut butter. If we don't have a symbol for it, it's harder to perceive and harder to remember.

Literate Chinese speakers can look at seven Chinese characters and remember them pretty easily. People who aren't literate in Chinese will have a lot of trouble with that, because each stroke—and perhaps the specifics of the shape of each stroke—needs to be remembered separately, because symbols representing those characters as a whole do not exist in their minds. So some people look at a Chinese character and "see" a word. It's the person's interpretation of that ink on paper that makes the perception what it is. So don't think about perception as being a simple pulling of information from our sense organs. It's more like trying to find the right interpretation of the information our sense organs provide. These interpretations are based on our memories.

This is very clear when we try to make sense of images that are essentially meaningless. Images from kaleidoscopes are beautiful, but ultimately are devoid of much meaning. You might notice red here, a diamond shape there, but in general, it's just a lot of colorful noise. This is reflected in how poorly people can remember images from them. Anthony Wright wanted to run an experiment on pigeons and monkeys in which he exposed them to photos and investigated how well they could remember them. But when he ran a version of the experiment with people, he couldn't use regular travel photos because people remembered *all* the photos. The task was too easy, because people were so good at recognizing and remembering the things in them. So he used pictures from kaleidoscopes instead.[10] Even if you see the same kaleidoscope picture later, it's hard to tell if you've seen it before or not, because no large patterns jump out at you. There's little for your mind to hold on to.

Another reason our minds need to create high-level symbols is because they need to be used by a wide variety of systems in the mind, not just vision.[11] To take a simple example, suppose you look in the cupboard and see peanut butter. Later, you are asked if there is peanut butter in the cupboard. The question is posed in language, but the experience you use to answer the question was visual. There has got to be some way for your mind to connect the sounds of the words "peanut butter" to the image of peanut butter that you saw. It does this by making a symbol that is independent of any particular sensory system. Your visual system is a machine that turns light into symbols.

It's important to distinguish the symbols we use to do general reasoning from actual words that we use in language. Each word is represented by a complex collection of symbols, but not every symbol our mind uses has a corresponding word. Different languages attend to different aspects of visual scenes, and speakers of different languages attend to different aspects of scenes, though there are some cross-language similarities.[12] But there are lots of visual symbols we don't have words for. For example, although we can distinguish the faces of Audrey Hepburn and Natalie Portman with ease, we would have a very hard time describing in words what those differences are.[13] Experts create their own symbols and jargon to describe things the rest of us don't—or can't. One time, I went clubbing with a bunch of dental students. Outside the club, they talked about the various cute guys they saw in the club. At one point there was confusion about exactly which guy they were talking about, and someone resolved it by describing the characteristics of his incisors. "Oh yeah, him," the other women said.

How does the symbolic nature of perception and memory affect our imagination? We tend to populate our imaginings with symbols, things we know and have experience with. The symbols we have in our memories affect the imaginings we are able to make. Creative imaginings are new, interesting combinations of these symbols.

When called upon to imagine a particular symbol, like a bird, and we picture one in our heads, what do we do? There are two basic possibilities: we might bring to mind some specific instance of a bird we saw at some point in our past, or we also might imagine a sort of average bird.

So picture a bird, if you can (I know not everybody can). What does it look like?

Most people picture a small songbird, rather than some unusual bird like an ostrich or an emu. One theory of why we do this is called "prototype theory." It holds that for every concept, as we experience instances of that concept (for example, each new bird we see), we slightly modify an averaged representation called a "prototype." It claims that we all have a prototypical bird somewhere in memory, which is the average size, color, and has the average beak length of every bird we've ever experienced.[14]

This theory is controversial. But whatever the underlying structure of our concepts are, it seems pretty clear that when people imagine objects and scenes, they behave *as though* they had a prototype in mind. Think of a birthday party. For most people, the birthday party they bring to mind has kids, presents, cake, and maybe games. People don't tend to think of four guys celebrating a birthday at a sports bar, shooting tequila. But that's a birthday party as much as a Himalayan goldenbacked three-toed woodpecker is a bird.

So when we think about how imagination works, we need to think about prototypes. Our memories are structured in such a way, or are accessed in such a way, that people will tend to populate their imagined scenes with prototypical instances of objects. They don't think of weird examples of things unless they have a specific reason to do it.

We can also distort shapes in our memory to be more prototypical—like the autocorrect I mentioned before. For example, if we look at the shape of a river on a map, unless the particular shape of it has a special meaning to us, we'll store it in memory as being more of a simple straight line or a curve.[15]

I've been talking about prototypes in terms of visual features, but it's likely that we also have prototypes of spatial relationships. When we think of a restaurant, we can bring to mind a generic place, and have an idea of where all of the important elements would be placed.

SPATIAL REPRESENTATION

One of the fascinating things about memory is that the mind uses more than one representation of space. Just like you might give someone directions with a map, or with left-and-right turn directions based on landmarks, the mind itself has several ways of remembering where things are. We even seem to have specialized systems for remembering the locations of our own body parts.[16]

When we look at a scene, we pull out spatial relationships between the objects in it, mostly without even having to try.[17] When we remember a scene, we can remember it as though we were in it. But there are a couple of ways our minds might pull this trick off: perhaps we remember all of the things that were around us equally well, no matter where they are in relation to us, as we might if our minds represented everything in a viewer-independent coordinate system. So, for example, the things behind us would be recalled (in our imagination) just as well as the things in front of us. Or, perhaps we store something like a picture of the scene remembered, as though we were actually looking at it (which would exclude certain parts of the scene, such as the things behind us). Or, perhaps we tag the location of objects in the scene according to where they are *in relation to us*, rather than in some more objective location (the jar of peanut butter is to my right, and the door is behind me).

Nancy Franklin ran an experiment to see which one of these was right. The experimenters described scenes (in words) to the participants. For example, the participants might hear, among other things, "in front of you is a plaque." Then they asked the participants to, in their imaginations, turn 90 degrees to the right. Then they were asked about the locations of the objects described.

If no spatial locations were privileged, then they should have been able to respond with object locations equally fast, no matter where the object was in relation to their imagined self—whether it was behind them, above them, or right in front of them. But that's not what happened.

If they were using a mental picture (of what would be seen from a particular point of view) to represent the scene, then they should be faster when responding about objects that would be visible—for example, they should be slower for things behind them, because those objects would not be visible in the picture. That didn't happen either.

What happened was that they were slowest at describing things that were either on the right or the left, a bit faster for things in front of them versus behind them, and fastest for things above or below them. What the—?

According to Franklin and Tversky, when we do an exercise like this, we are coding locations according to three axes: up-down, front-back, and left-right. The speed at which we are able to retrieve locations from memory depends on which axis is most important. Some axes are easier to distinguish than others. Our world has a strong up-down asymmetry because of gravity. Things that are up in the air tend to fall, they tend to be smaller, and so on. The fact that things up high behave differently means that it's very important to be able to distinguish between what is up and what is down. This is why the locations of things either above or below are identified the fastest.

The next most important asymmetry is front-back. This asymmetry is personal. You can see and more easily interact with things in front of you. What's in front of you changes as you turn around, but what's above and below you does not. This asymmetry makes it the second fastest.

Finally, we have the left-right axis, which changes as you rotate, but has no other significant asymmetries (aside from the direction we read and write—we'll get to that later), and as such, it's less important, and thus, more difficult and slow to distinguish things on the left from the right.[18]

So if we were remembering objects in 3-D space according to some objective, three-dimensional coordinate space, we'd be just as fast retrieving everything, no matter where it was in our imagination. If we were using a mental *picture*, we'd be slower at identifying things behind us than things left and right, but that doesn't happen either. It appears that we place objects in locations according to where they are in relation to our bodies.

Or, at least, that's one way we do it.

Franklin and Tversky also found that if the scene is described from the perspectives of two different people in the same scene, people would show no response time differences at all, suggesting that people can also create a 3-D, perspective-neutral version of the scene! It looks like people represent scenes in memory differently, as needed for different tasks.

We also seem to represent locations in terms of continuous space, but also in terms of categories for general locations. Jannelen Huttenlocher had people study the locations of dots on a screen, and then asked them to remember, later, where they were. She found that people got the locations subtly wrong—they drifted toward the center of the categorical areas in the visual field: top, bottom, left, and right. It's as though, at some level, we think "that dot is on the right," and then put it closer to the center of where we think of as the right side like an auto-correct.[19]

Thus, it appears that we use place categories for the locations of objects. But how do we represent the relationship between two objects? In terms of coordinates, or simple relations like "above" and "next to," these questions are very difficult to answer, because the mind creates new representations as needed. Sometimes, when people are asked about something they remember, they give answers that suggest that they have a categorical, relations-based memory of space, but at the same time, they can reproduce a map they memorized, as though the quantitative distances between objects were memorized. The answer seems to be that we understand space in many ways and can, on the fly, transform that understanding into other formats as needed. Some of these even have different neural signatures when studied with brain imaging.[20] It's a powerful, wonderful thing, but it makes it fiendishly difficult to study in experiments because you can't always know (let alone control) what people are or are not doing in their heads.[21] When people behave differently in different situations, it could be that they are relying on fundamentally different kinds of memories and even creating new representations on the fly.

These representations of space we have are also distorted in interesting ways. For example, most people think of Europe as being east of the United States—which, of course, it is. But that's only the most obvious directional difference. What's easier to forget is that a lot of Europe is also *north* of the United States. But in our memories, the north-south axis is not the first thing that comes to mind, so we tend to align the continents in our minds along the east-west axis. So people might think that Berlin, which is in central Europe, is at the same latitude as the central United States. But its latitude actually crosses Canada. Los Angeles and New Orleans are on the same latitude as Northern Africa.

The same goes for North and South America, which we think of as a lying in a fundamentally north-south relationship. We tend to think of North America being more directly north of South America than it actually is, forgetting that South America is also quite a bit more east.[22] The westernmost tip of South America is actually below Florida.

We also get distortions according to what we know well versus less well. We tend to think of places close to us as being larger than places far away.[23] For example, we might think of the cities in our own country as being more distant from one another than cities in another country are from one another, because we think of foreign cities as, fundamentally, being in the same place—that foreign country. Sometimes my American friends will talk about coming to Canada and visiting the places they've heard of, only to be surprised to find out that the distances between cities in Canada can be very, very vast. But without familiarity with the actual geography, their minds condense the information: Calgary and Ottawa are both in Canada, so they must be close to each other (they're not; it's about a four-hour flight).

When we don't know the details of a space, we compress it in our memories. For example, you might be able to think of many more things between your home and your grocery store than between Rome and Florence. As such, there's less to remember in the latter, and your mind can superficially interpret this as being less distant than it really is. When people sketch maps, they compress distances between places that have no turns. That is, even if a road is really long, if you have no turns to make, they will draw it as a shorter line.[24] One time my beloved and I walked through a dense, crowded, interesting place in London—Borough Market. We decided at some point to go back to a restaurant we'd passed and were surprised to find that it was only about forty feet behind us! But because we'd seen so much in those forty feet, if felt like it was a quarter of a mile away.

This is also why return trips feel like they are shorter, in terms of time and distance, than trips there. We are already familiar with the things we saw on the way there, and so our mind compresses them, making it feel like a shorter distance and time.

MEMORY FROM TEXT

I've been talking mostly about how we make memories from things we experience in the world through visual perception. But we also learn from language. Of course, when we hear someone talking about whether Marilyn Monroe was prettier than Anna Nicole Smith, we are using our auditory perception, and when we read text, we are using visual perception, but in these cases the important thing is that we're understanding language, and it's less important whether we got the words through our eyes or our ears. As such, the specifics of the visual and auditory systems fade in importance.

In fact, people often can't remember whether they heard or read something, and have a hard time remembering exactly what words were even said. People remember what's important: the meaning. When a scene is described to you, you create a "situation model" to help you comprehend the sentences. It contains a mixture of what you already know and the meaning of the sentences. The situation model includes all kinds of information: when events happened, how things are spatially organized, the causal relationships between the elements of the situation, and even the goals, attention, and other mental states of the characters. For example, if I tell you that I was riding my bike, and I tapped on the window of a car to talk to the driver, you will create a situation model that includes all of these things: *I had intended to get the driver's attention. The tapping made a sound. We were both on the road.* I didn't mention any of those things. They are brought to bear by your mind as relevant information. And later, you are unlikely to remember what was in the sentences versus what you inferred from them.[25]

Clever experiments by psychologists have found that these situation models are organized by locations. Mike Rinck gave people some text to study that described a bunch of objects in a bunch of locations. Later, they were asked about what objects were where. If they were asked about several objects in the same location, they were fine. But if they were asked about the same objects in multiple locations, they slowed down significantly. What we think is happening here is that the remembered meaning of the text was organized according to the locations, and each location was populated by a number of objects.

Think of it like a filing system. The mind creates a file folder for each location and puts a file in each folder for each object. What people don't seem to be doing is making a folder for the objects and putting locations in them. If I'm asking you about the contents of one folder, it takes less time than if I ask you about things in multiple folders.[26] You can probably relate to this yourself: it's easier for you to list the objects in your living room than to list all of the locations you've ever seen a plush dog toy.

3-D OR 2-D?

I've been talking a lot about how we remember the locations of objects in scenes. But visual memories also have information about the objects themselves—most objects are complex, consisting of parts. A tree, for example, has a trunk, roots, branches, bark, and leaves, and we know not only that trees have these parts, but the spatial relationships between them: the trunk runs up through the middle, and leaves are attached to branches. An interesting question is whether our visual memories of the parts of objects are stored as two-dimensional pictures or as three-dimensional structures.

This question is full of subtleties that must be unpacked. As I discussed in the first chapter, we're pretty sure that we don't store "pictures" in long-term memory at all—they are symbolic structures that could be used to create pictures. But if the things in memory are symbolic structures, not pictures, then what sense does it make to ask if they are stored in two or three dimensions?

Let's take a television as an example of how we might represent its parts symbolically. We might have symbols describing its shape (rectangular), the fact that there are six little round buttons on the bottom right, and a platform beneath the screen to hold it up. This is a frontal view of the television. It has no information about what's on the back, such as the cord, or the various input ports. This is what it might look like to have a symbolic description of two-dimensional information.

Research by Michael Tarr suggests that people's memories of objects are collections of two-dimensional views from various angles—like how someone might photograph a sculpture from various angles to sell it online. Some of these viewpoints are preferred over others. Let's take someone's concept of a chair. If they had a true 3-D representation of the chair in mind, then they should be just as good at recognizing and imagining a chair from any orientation. Front, back, from the top, etc. But that's not what happens. If you show people the view of a chair from the front, with the legs on the bottom, they are faster to recognize it. This supports the idea that we store our visual memories of objects as a collection of 2-D views.[27] But they are *symbolic* descriptions of 2-D views, not pictures made of pixels.

But given the flexibility we see in how people can represent objects in scenes, it would not surprise me if we eventually found that people represented complex objects in multiple ways, too—and indeed, many psychologists, such as Irving Biederman, believe we store complex objects as collections of primitive 3-D shapes.

So it appears that we look at scenes, remember them in possibly multiple ways, and can change them in our working memory. But what about people who are blind from birth? How do they understand scenes if they can't even see?

Blind people get asked all the time what they "see." By this, people want to know what blind persons' visual experiences are like. The answer might seem obvious—that they see nothing but blackness, as though a perfectly dark curtain had closed over their visual field. This seems obvious because this is what we experience when there's no light. But for people who have been blind from birth, they don't see blackness, they see something different: *nothing.*[28]

What's the difference? Isn't darkness the same thing as nothing, simply the absence of light? In physics, perhaps, but in terms of mental experience black is a color like any other, and seeing any color, including, black, would be seeing *something*.

So what does it mean to see nothing?

To understand this, let's consider sensory systems human beings *don't* have. Some animals, such as geese and loggerhead turtles, can sense the Earth's magnetic field and have a natural sense of direction. That is, they can feel which way is north as easily as we know up from down. Now imagine for a moment that a goose was smart enough to talk to, and you explain to the goose that you, like all your fellow humans, don't have a sense of the magnetic field at all. The goose is confused, and perhaps thinks of a time when she went to a goose science museum and they had a special room that blocked the magnetic field of the Earth. When she went in there she got the sense of what it is like to be in no magnetic field. Perhaps geese even had a word for the feeling of being in no magnetic field: *mlack* ("magnetic black").

So naturally, the goose asks, "So you humans sense *mlack* all the time?"

No, you'd explain, you don't sense *mlack,* you really have no idea what *mlack* even feels like, because without ever having had any sense of magnetic fields, it feels like nothing in particular when there is no magnetic field. You sense "nothing" when you try to detect the Earth's magnetic field. You don't sense *mlack*, and blind people don't sense black.

Even fully sighted people can get a sense of a visual "nothing." Look straight ahead. Note that your visual field has a boundary. If you spread your arms, drawing your hands back until they are just out of view, you notice the limits of what you can see. Now what does your hand, or anything else, "look like" beyond this boundary? What do the things behind you look like? Do they look black? Nope. They look like nothing at all. That's the difference.

Some people have an interesting disorder called "hemispatial neglect." They can't see, and so ignore, one half of their visual field—either the left or the right. When they eat dinner, unless they remember to rotate their plate around, they will only eat only one half of it. But they don't see a black blob on the neglected side, it's simply a different boundary on their visual field. They see nothing there.

It's interesting to think about the imagery abilities of the blind. Congenitally blind people, like aphantasics, lack visual imagery, but they remember spatial information, and everything else, just fine.

EPISODIC MEMORY

Most of us can think of things that happened in our past and can picture them. Psychologists call one's memories of things that have happened to them "episodic." This is to distinguish them from memories of certain facts that might not be tied to a specific event, such as your knowledge of the order of letters in the alphabet. You have in your memory the letter order, but can't recall a specific where or when you learned it. More abstract knowledge like this, or the fact that the Beastie Boys are ill, is called "semantic memory." Another kind of memory allows you to walk and ride a bike—these memories of how you do things are called "procedural memories." Very young children can remember the *semantic* content of events, but the ability to remember past *episodes* emerges at around age four.[29] In fact, until that age, children cannot even mentally consider a state that they are not currently experiencing.[30]

But as we've learned, our long-term memories are made up of symbols, not pictures. When we recall these sets of symbols, they get reconstituted and filled in with information from other memories. Then, if they are imaged, they get fleshed out further with other information that wasn't from that memory at all, but from memories of other things. Because the images you create from episodic memories are reconstructions, the process is prone to errors, and ends up being slightly different every time a particular memory is recalled.

Recalling an episode from your life might be better described as an episodic reconstruction—bits of true episodic content liberally fleshed out with semantic knowledge.[31] For example, suppose Fatima remembers eating dinner with her parents, and pictures the scene in her head, experiencing the memory vividly.

Suppose that when she actually ate dinner with them, she never looked up at the ceiling. What this means is that what the ceiling looked like *at the time of the event* is not stored with that episode. Nevertheless, a week or so later, when she imagines the dinner, she can "look upward" in her mind's eye and see the ceiling. But if she hadn't paid attention to the ceiling at the time, she wouldn't have encoded an episodic memory of the ceiling in the first place. There might have been a big spider up there, but that spider would not be in her memory of the ceiling. Her "memory" of the ceiling would be that it was empty, as it normally is. How is she able to "remember" it, then, if she didn't look at it? Why does her mind put an empty ceiling in there?

Because she *knows* that the room in question has a ceiling, and she might even know from other episodic memories what that ceiling looks like, from a time, perhaps, when she was lying on the floor. So when her imagination "asks" what is above the table, semantic knowledge slips in the information she needs from other memories.

The interesting thing is that we don't even know when we're doing this. In Fatima's case, looking up at the ceiling in her mind's eye feels natural, and the look of the ceiling doesn't feel particularly different from the memories she brought to bear from the actual episodic memory. It happens so effortlessly and quickly that she experiences the illusion as if it was stored like that in her head the whole time. Her mind fills in the blanks so effortlessly, that she can't even tell what is a part of the actual episode and what is missing.[32]

In other words, each act of remembering something from your past (episodic memory) is also an act of imagination. Our memories hold clues to what happened in our past, but some of it is completely make-believe. If this is true, then episodic memory recall and imagination should both be good or bad in the same person. And, indeed, Daniel Schacter shows that people who have trouble with one (imagination or episodic memory recall) tend to have trouble with the other, and that imagination and memory use similar brain networks.

If many acts of regular old memory recall involve elaboration with things from other memories, then putting a clear boundary between what's imagination and what's not is impossible.

Because our memories are *mostly* right, scientists like to focus on when our memories get things "wrong." And what does "wrong" mean? That's when our recalled beliefs are not true. One of the ways we can remember things incorrectly is described in the literature called "false memories."

FALSE MEMORY

The simplest way to demonstrate false memories in the laboratory is by asking people to remember a list of words that are all related to a particular topic. Suppose you ask someone to read this list of words: bed, rest, awake, tired, dream, wake, snooze, blanket, doze, slumber, snore, nap, yawn, peace, drowsy. When you ask someone later if the word "sleep" was on the list, they are more likely to say "yes" than if you ask them if the word "see" is on the list. Recall that when we hear somebody say something, we remember the meaning of what was said and have a hard time remembering the exact sentences. It appears that our memory works similarly with lists of words: we remember the gist of the list, and have a hard time remembering specific words.[33]

Sometimes we can tell a memory is wrong because we use other memories to cast doubt on them. For example, I have a memory of riding my tricycle down the stairs. On the one hand, the memory is real, because when I was a kid, I did just that. The problem is that in my version of the memory, in the little video that plays out in my head, it happens *in the wrong house.* I lived in two houses when I was a young child. The first one I have no memory of because I was too young. So I remember the tricycle incident incorrectly, in the house I lived in later.

How can this happen when the memory in my head seems clear as day? It turns out that I rode my tricycle down the stairs before I was three years of age, and most people can't remember much that happened before they were three. I have no actual memories of the house I lived in then. But the story of my riding down the stairs was told over and over in my family. Many families have legends, stories they tell again and again. Every time the tricycle story got told, I *pictured* it happening in my mind. Vividly imagining things happening to you can sometimes result in you mistaking them for memories of things that actually happened. Now this is one of my memories, still there even though I know that it's substantially wrong.

False memory has some very serious real-world implications—even more important than kids tumbling down stairs on tricycles. A woman named Elizabeth Carlson was admitted to a hospital with symptoms of depression. She was told that she suffered from multiple-personality disorder (this is now known as dissociative identity disorder, and just about everything about it, including its very existence and treatment, is controversial in scientific psychiatry). Her psychiatrist told her that she had repressed memories of childhood abuse in addition to the ones she could remember. The psychiatrist had Elizabeth engage in three therapeutic techniques to try to release these memories.

The first was bibliotherapy, which involves reading accounts of other patients (some factual, some fictional). She was told that if her body felt irritated while reading these accounts, that it was further evidence of her multiple personalities.

The second was the use of guided imagery in which the therapist helps the patient flesh out fragments of memories, and even hypothetical situations, into fuller episodic memories from her childhood. Elizabeth acted out episodes of molestation during therapy. Her psychiatrist told her that her nightmares were recovered memories.

Finally, Elizabeth engaged in journaling. The psychiatrist interpreted these written thoughts and feelings in a context of childhood sexual abuse.

At this point in the book, you probably know enough about imagination to see the train wreck that was coming. Elizabeth started "recovering" horrific memories from her childhood of satanic rituals, many conducted by her parents. She even "remembered" eating the flesh of a dead baby. She started discovering the names of her other personalities. She suspected her parents were satanic cultists who wanted to murder her and were trying to control her with posthypnotic suggestion.

As her therapy went on, her condition didn't get better. It got worse. She was heavily medicated, slept most of the day, was fearful to leave the house, often suspected her husband of wanting to murder her, and left the care of her family to one of her children, who was still in high school.

Then some of the patients in the hospital started meeting on their own. They noticed something fishy going on. Several of them were diagnosed as having multiple personalities, but, strangely, many patients found that their personalities had the same names as other patients' personalities. They shared their "memories" of sexual abuse and found that they were very similar—and, even more disturbingly, very similar to the accounts they'd read in bibliotherapy. The patients discovered that what they and their psychiatrists had taken to be recovered, repressed memories *never actually happened at all,* even though they felt as real and vivid as other memories.

Elizabeth and another patient took the psychiatrist to court and were awarded two million dollars in damages. Their argument was that repressed memories don't exist, and that the memory-recovery therapies were bogus.[34]

The lesson of the Elizabeth Carlson story is that vividly imagining something happening to you can result in your inability to distinguish it from an actual memory. These imaginings can come from stories told by others, leading questions from police or therapists, guided visualization, hypnosis, or even literature, movies, and television. When a client tells a therapist about events that he believes happened to him, the therapist tends to believe him, and even false memories can be reported with very moving and convincing terror, rage, depression, and guilt.[35]

Maryanne Garry ran a study in which she described to people a list of events that might have plausibly happened to them in their childhood, such as breaking a window with their hand or falling down the stairs. She also asked them to rate how confident they were that these events actually happened. Then she had them imagine some of these events happening, but not others. Sometime later, the people were more confident that the imagined scenes had actually happened than the ones they hadn't been asked to imagine. This process is known as "imagination inflation."[36]

The theory behind this is that we use the ease with which we can bring images to mind as indicators of many things, including how common the situation is in general, how probable similar events are to happen again, and, indeed, whether they actually happened at all. This ease is called "fluency." What can happen is that people mistake fluency for familiarity, clarity, liking, fame, and even beauty.[37]

Elizabeth Loftus and her laboratory have run a number of experiments that managed to implant false memories of events into participants simply by encouraging them to imagine them happening. For example, in one famous study, she told people about several childhood events that had been recounted to them by their parents. All of the events in the list actually occurred (as far as the parents' memories can be trusted, anyway) except one: a story about being lost in a mall and being helped by an elderly woman before being reunited with their family. A full 25 percent of the participants in this study came to believe that this actually happened to them, and many provided embellishments, such as details about what the elderly woman was wearing! Why does this happen? Because when we imagine something, it becomes more familiar, and when familiar things feel true, we forget that imagination might have caused the familiarity.[38]

As we saw with the story of Elizabeth Carlson, this finding has important implications for law. If well-meaning therapists and police officers can implant memories without even meaning to, it calls into question the quality of eyewitness testimony in court. We tend to be more convinced of the truth of a recounted event if it is told with conviction and has lots of details. But Loftus's studies show that both of these things can happen with "memories" of things that never happened.

This calls into question the existence of memory repression at all. One study found that of the sixteen children studied who had witnessed one of their parents being murdered, not a single one repressed the memory. As is common with traumatic events, whether abuse in the home or horrific experiences from war, as the children grow up, they couldn't help but have emotion and preoccupation associated with the event.[39] What we see isn't that people repress traumatic memories, the effect is quite the opposite: people can't stop thinking about them, and that's often why they seek medical help.

Naturally, there was a backlash. The idea that people can repress memories is commonly believed, by therapists and the general public alike, and is one of the foundations of Freudian psychoanalysis. Loftus's critics suggested that being lost in a mall is quite common, and it could be that 25 percent of the study participants actually *were* lost in a mall. Perhaps the experiment was helping them access memories of things that had actually happened to them.

In response, Loftus followed up the study with another one. She suggested to people that they had met Bugs Bunny at a Disney resort. This was accomplished, in part, by showing them fake advertisements for Disney featuring Bugs. This led 16 percent of the participants to believe they had personally met Bugs at a Disney park. Of those, 62 percent said they shook his hand, and 46 percent reported hugging him. This finding is significant because Bugs Bunny is a Warner Bros. character and would never appear at a Disney resort. The researchers had managed to convince people that an impossible event actually happened to them. Guided imagination, stories from other people, and suggestive feedback can give people rich, false memories, and people with more vivid imagery are more susceptible to it than everybody else.[40]

False memories can also come about as a result of repeated questioning in a legal context. Every time you recall a memory, it is subtly altered, and your imagination fleshes it out and changes it according to the context in which it's retrieved. Repeated recall of the fleshed-out, imagined parts can become indistinguishable from memories of what you actually experienced in the past.

False memory can be terribly abused in police interrogations, particularly because police are allowed to lie to you during questioning. Sometimes witnesses and suspects end up admitting to things they didn't do, or saying they saw things they didn't see, based on ideas introduced by the interrogator. For example, a witness might be asked repeatedly if they saw a man with bushy eyebrows. As the witness searches their memory for such a thing, each time they do it they start to associate whatever memory they actually have with bushy eyebrows more and more. By the time the interrogation is over, the witness might be convinced that the cop helped them remember something they saw, when in reality they have created a false memory by using guided imagery.[41]

You can see the problem. Freud introduced the idea that we can repress memories and that by repeatedly trying to recall them we can uncover what really happened to us. This is a part of the folk psychology that everybody believes in. The idea that your imagination can create false memories is genuinely surprising to people. So when people's reports of what they remember change over time, or during the course of an interrogation, police, witnesses, suspects, lawyers, judges, and everybody else rarely think of false memory as an explanation. From my reading of the literature, however, it seems much more likely that the answers given to questions asked earlier in the investigation will be closer to the truth than the later ones, because the later ones will inevitably be populated with characteristics that were merely imagined during repeated questioning.

It's clear that false memory can cause a lot of problems. But a positive side is that you can use your imagination to interfere with your mind's ability to store things that *you don't want to remember*!

One view of the mind is that we have a certain amount of mental energy, and if you use mental energy doing one thing, it "starves" processes that are trying to do another. Suppose you witness a terrible car accident, and see blood and injured people, and you're worried about it keeping you up at night or giving you nightmares. You can interfere with remembering the scene by otherwise occupying your mind just after the event happens. Emily Jones ran an experiment in which she had people watch a traumatic film featuring injury and death. Then she split this group in two. The part of the group that played Tetris for a half hour afterward had fewer "flashbacks" over the course of the next week. The theory behind this is that traumatic experiences are initially encoded as sensory experiences. The visual stimulation from playing Tetris immediately afterward interfered with the mind's ability to process the visual information of the film. They remembered the traumatic stimuli less and, as such, were not as bothered by it in the future.[42]

Post-traumatic stress disorder (PTSD) is a problem, in part, because it forces you to imagine unpleasant things in flashbacks. It appears, though, that occupying the visual areas of your brain, either through imagination or computer games, can reduce the effect of PTSD if used right after the traumatic event. So if you don't want nightmares after you watch a scary movie, play some Tetris.

But if you don't have a video game handy, you can do it by imagining something very vividly. In a study by Cheryl Rusting, people in an angry mood were asked to vividly imagine, for example, riding around a city in a double-decker bus. The imaginers had decreased feelings of anger compared to people allowed to ruminate about what made them angry.[43]

So it's clear that we can create false memories, and that even our relatively accurate memory recollections are fleshed out with imagination. But for the most part, we are pretty good at knowing the difference between what we imagine and what we remember.

HEAD TO THE SOURCE

Imagine being in a grocery store on a Mars colony. You sneeze, and the grocer says, "Sounds like somebody's allergic to great prices!" A few days later when you recall imagining that, you will probably know that it was just imagined, no matter how vivid your imagination is. That is, you won't mistake it for something that really happened to you. How are you able to do this? Given that imagination and actual perception use a lot of the same perceptual brain areas, and that imagining can lead to false memories, how do you ever know if a memory really happened or you just imagined it?

Your ability to distinguish where memories come from is known as "source monitoring," because you're attributing the source of something in your memory. That is, should this memory be attributed to an experience in your past, or an imagined experience, or something somebody told you?

One idea is that when you create a memory, you tag it somehow, in your mind, with what kind of memory it is. Was it something that happened to you, or something somebody told you, or was it a fantasy, or a dream, or a plan for the future? According to Steven Lindsay, this "tagging" doesn't actually happen, at least not at first. He claims that the source is inferred from the information in the memory itself. Real episodic memories, for example, tend to be more vivid and have specific times associated with them. He says that when you retrieve a memory, you use cues like this to make a judgment, on the fly, about whether it really happened to you or not.[44] You will know that your memory of being in a grocery store on Mars never happened, in part, because you know you've never been to a Mars colony.

Getting the source wrong can be mildly irritating. I know that I often recall some "fact," but then cannot remember (or, often, track down in my notes) where on earth I heard it. I just have a vague notion that I had "read it somewhere." It gets more serious when you start mistaking imagination for memory, or mistaking something in a dream for something that really happened, or, in extreme cases, you take delusions and hallucinations as real.

People use a variety of ways to evaluate sources. The vividness of the recall is an important one. We might feel confident that we saw something happen because we remember specific details of the scene in question. We even use the memory of details to try to convince other people that we actually experienced it: "I remember she was sitting right there when she said it." This is a part of why vivid imagination is more likely to lead to false memories. The memory of the imagining is detailed and vivid, so it feels more real. People who have vivid mental imagery tend to do better at visual tasks, but they actually do worse at source monitoring—the vividness of their imagined scenes is so convincing that they are more likely to mistake their own imaginations for real memories![45]

We can also adjust how stringent we are. In casual conversation, we might be happy to attribute a flattering idea we have about ourselves to something that really happened, but might be much more circumspect when testifying under oath in court.[46]

Imagine that you watched a horror movie while in a cabin in the woods. If, later that night, you thought you heard footsteps outside, you might wisely attribute this perception to imagination, because you know you were scared and listening for any sound that might be interpreted as something dangerous. Your guard is up, so to speak, because you know the horror movie rattled you.

All of this is important for understanding how imagination works, because your mental images are not much different from your perceptions. But the problem goes deeper.

For example, imagine that you could fly like Peter Pan. Imagine being on the roof of a building you know well and flying off of it. Look down. What does it look like? If you were to travel to a nearby location, it might take you less time than if you had to move on the ground, because in the air you have fewer obstacles. What might that trip look like from the air, flying just above the buildings, in a straight line to your destination?

Now think about the reasoning you did to come up with that answer. When you engaged in this exercise, you were engaging in hypothetical reasoning and possibly some mental imagery, too. You were *supposing* that you could fly. But, of course, you can't really fly. It was a hypothetical, make-believe situation. But also notice that you could effectively reason about what the experience *would* be like. When thinking about a direct route to another location, you used the same kind of reasoning you'd have used if you were thinking about the path a bird might take and what a bird might see—something that *is* possible. You were using the same kind of reasoning to think about possible and impossible situations.

I just asked you to imagine that you could fly. But you're not going to go jump off of a building, because you used source monitoring to know that the idea in your head is one generated by your imagination, and not some true thing you learned about the world. We don't go running from the house when we imagine that it's on fire. Even when a kid is pretending that a banana is a phone, they don't expect to hear people talking through it.[47]

If you *actually* could fly, you'd probably use the exact same kind of reasoning to plan your Saturday, and how you would zip around town, running errands. What this indicates is that you use the same reasoning processes whether you are reasoning about things you really believe or things you are just supposing. You also use this when you decide that a movie or book plot doesn't make any sense—you are using your reasoning abilities on hypothetical information and coming up with conclusions.

So your hypothetical situations have their own conclusions, which also have to be kept separate from actual, real-world conclusions! In other words, the things you conclude about a make-believe situation are also make-believe, and your mind needs to know the difference.

Because you can use the same reasoning processes, it makes sense to suppose that hypothetical suppositions and actual beliefs are encoded in your mind in the same kind of format. You might have a reasoning system, for example, for thinking about how a crowd of people might behave when frightened, and you apply this same system whether you are thinking about the crowd you're currently in when the fire alarm goes off or when you are reasoning about the behavior of characters in a similar situation in a novel. Just as a file on your computer needs to be formatted in a certain way for your word processor to open it, the beliefs and hypotheticals need to be in the same format so that your reasoning processes can work with them.[48]

Further evidence that they are the same kind of thing comes from our emotional reactions to imaginings. We might cry watching a sad movie like *Finding Nemo* and might get upset if we vividly imagine the death of someone we love. So, our emotions, as well as our thoughts, respond similarly to imagined facts and to actual facts. Indeed, many people actually consume fiction with the explicit goal of managing their moods.[49] This demonstrates what is known as the "paradox of fiction": Why would we feel emotional about things we know to be completely made-up? We can sometimes make a horror movie less scary by telling ourselves that it's just a movie, but the remarkable thing is that movies can manage to scare us at all, when we know full well that it's just light on a screen depicting fictional characters created by writers and designers.[50]

The source monitoring theory would say that we evaluate on the fly whether a particular memory really happened, based on things like vividness. But conclusions drawn from memories are never actually experienced in the same way we experience things actually happening to us. So vividness probably doesn't play a large role for these conclusions. How, then, are we able distinguish conclusions drawn from things we actually believe from conclusions drawn from imagined situations, given that they are encoded the same way, probably have similar levels of vividness, and use the same reasoning processes?

As I write this, I am sitting in a house. I can imagine that if there were Godzilla-sized creatures in the world, that this house might be vulnerable to being crushed by them. I drew that conclusion from "beliefs" about Godzilla, and combined them with real beliefs about the house I'm in. In this instance of reasoning, I'm using a combination of reality and fiction. My conclusion that this house is vulnerable to giant monsters is clearly not a part of reality, because I don't really believe in giant monsters. How do I know this? Well, an important part of the reasoning involved an imagined "fact," not a real belief, which means that the conclusion drawn is also mere imagination. So my mind classifies the idea that "this house is vulnerable to giant monster attacks" as a belief-like entity of an imagined situation, not an actual state of the world.

In cognitive science, we use a metaphor of infectious disease and wonder why your hypothetical beliefs don't "infect" your real beliefs, leading you to incorrect, and potentially disastrous, effects in the real world like jumping off of buildings simply because you imagined you could fly. Indeed, we seem to have no trouble keeping our pretend beliefs separate most of the time (with dreams, delusions, and many hallucinations being counterexamples). Even children engaged in pretend play know the difference. If a child is pretending that a dish scrubber is a hot dog, she has, what we might call, "double knowledge": in the context of the game it's food, but she won't try to actually eat it.

Philosopher Shaun Nichols suggested that what we do is put these "make-beliefs" in a "pretense box." This box keeps the make-beliefs quarantined so that inferences using them don't infect the real beliefs we have. So when we're reading a mystery novel, we might figure out who we think the killer is. This make-belief, say, that Pat is the killer, stays in the box. Then we don't go around thinking that there is some real person named Pat who is a murderer.[51]

The idea of the pretense box is very different from the source monitoring theory I talked about before. Do we know ideas are fictions because they are in a special "box" or are somehow tagged as fictions, or do we infer that they are fictions on the fly because they are less vivid (as Stephen Lindsay's theory would suggest)?

Probably both are happening. When we first attribute a source to a belief, we use the source monitoring processes to infer where that belief came from—and with that, whether it's true in the real world or in some fictional or hypothetical one. But the very act of inferring the source generates a new memory *of the source*, that can then be retrieved like any other memory in the future. That is, when you think of the time you imagined flying, you conclude that it must have been an act of imagination for various reasons, including your knowledge of the real world that says that people can't fly, the lack of vividness and detail in your memory, and so on. This is source monitoring. But once you have attributed it to imagination, you can store a new fact in your memory saying so. When you need to know whether it was an imagining or not sometime in the future, you can simply recall the newer memory, or the tag that says it's only imagination. You don't have to look at the evidence all over again and recalculate. This tagging of beliefs and make-beliefs in memory, according to their sources, might be how the "pretense box" works.

If there are only make-beliefs in the box, then how can you reason about them? The answer is that real beliefs from outside the box can affect make-beliefs inside the box, but not vice versa. So you can use your real-world knowledge that the grocery store is north of the Vietnamese restaurant to inform your hypothetical flight over the city and conclude new things that also go in the pretense box. But you can't use the imagining that you can fly to conclude actual things about the world, like concluding that it's a valid option for transportation to get to work tomorrow morning. There's an asymmetry there. Real-world beliefs can affect imaginary situation make-beliefs, but not the other way around.

When you read a novel, for example, you have to keep track of an enormous number of make-beliefs—imaginings you have to keep in mind that are true in the novel but not true in real life. Suppose you're reading Anthony Francis's urban fantasy novel *Frost Moon*. In it, there are vampires, werewolves, wizards, and magical tattoos. Each of these are slightly different from what you know about these things from other novels, and comprehension of them requires remembering and being able to reason about lots of make-beliefs that you accumulate over the weeks that you spend reading the book. These ideas, these imaginings, according to Nichols and Stich, would all go in the pretense box.

But can it be as simple as a single box? Do we simply tag some ideas as make-beliefs and others as true? The pretense box is an attractive idea, but can't be the whole story, as I argued in a paper I wrote with philosopher Jeanette Bicknell.[52]

Suppose you're reading the Lord of the Rings books, and it's taking you a few months. During that time, you watch one of the Chronicles of Narnia movies. After watching the movie, you continue reading the Lord of the Rings books.

According to the pretense box theory, your make-beliefs from the books are put in the pretense box or are tagged as pretense. This way you know not to believe that hobbits and ring wraiths exist in the real world. Same goes for the "facts," the make-beliefs, about Narnia. You know that Mr. Tumnus isn't a real creature, but that he's real in Narnia.

But you also know Mr. Tumnus won't show up in Middle-Earth (the world described in Lord of the Rings), and you don't expect Sauron will walk through his wardrobe and kill everyone in Narnia. We can all do this kind of segmentation effortlessly, but how?[53]

If the Middle-Earth and Narnia facts were all kept in the same box, or tagged simply as fiction, then we would not be able to distinguish those facts that are true for Middle-Earth from those of Narnia.

Somehow we tag facts as being true in certain contexts and not in others. So rather than just tagging the existence of hobbits as simply fiction or non-fiction, it's tagged as fiction that is true in the world of Middle-Earth, but not in Narnia.

What about the Shakespeare scholar who knows the plots and characters of all of Shakespeare's plays? She never makes the mistake of thinking that Othello will show up in *A Midsummer Night's Dream*, and she can remember these stories and keep them separate in her mind, for a lifetime. Let's call this the "many stories" problem. Not only do we have to keep stories quarantined from our actual beliefs, but we have to keep make-believe stories quarantined from each other!

But with just one box for imaginings, how can you keep one fantasy world separate from another? To extend the medical metaphor, having only one pretense box is like quarantining people with leprosy, polio, and ebola in the same room. For this reason, the explanation for how it works has got to be more complicated than a single pretense box.[54]

Our understanding of fiction gets even more amazing. Here's another example: suppose Julie reads the following very short story. When you read it, think about what kinds of make-beliefs Julie would put in her pretense box.

> David was doing his nightly prayers, facing the crucifix on the wall, when he noticed motion near the window. A bat had flown in, and was transforming into a human form, with a vicious smile and sharp fangs. "This can't be!" David thought, "Vampires don't exist!"[55]

In reading the vampire story, Julie might think that David should grab the crucifix on the wall. Why might she think this, when there is nothing in the story to suggest that vampires are harmed by them? She knows it from other stories she has read. She has an understanding of vampire lore, which she's collected over her lifetime of experiencing vampire stories.

This curious effect shows that although we keep stories separate from each other, there is some transfer. Julie can transfer her make-beliefs from other vampire stories to this one, because she sees them as a part of a class of stories—and she also knows that the Lord of the Rings is not a member of this class.

What's interesting about this is that most of us have heard many vampire stories—stories that don't happen in the same "worlds." That is, the vampires in *Twilight* are not in the same fictional reality as Count Dracula or the world of the movie *Fright Night.* So how is this crucifix fact tagged? There seems to be some kind of "vampire lore" tag, and all of the vampire stories (*Dracula*, *Fright Night*, and so on) are a part of that lore. In other words, the tags can be overlapping.

But she also might encounter a piece of vampire fiction in which it is made clear that crucifixes don't work. Then she believes that they don't work in *that* story, but that they work for vampire stories in general, all the while not believing in vampires in the real world. So the make-belief "crucifixes ward off vampires" is false in that story, true for vampire lore in general, and untrue in the real world. The same make-belief switches from true to false at each layer! This shows that our beliefs and make-beliefs are not simply stored as a matter of being true and false. They are true or false relative to particular contexts.

This is how we keep track of our day-to-day hypothetical situations, too. Suppose you are biking home and hear that a particular road is closed. You think of the possible alternate routes you can use. You could take one route, but there is heavier traffic there, or another one, but it's a bit longer. These hypothetical beliefs are not "believed" either—they are merely being entertained. We know they've got to be belief-like, though, because you can reason about them. They are *separate* hypothetical situations, so they must be tagged as such in memory so you don't think that you're taking the trafficky route and the longer route at the same time. Just like you need to keep Middle-Earth and Narnia separate, you need to keep your plans separate so you can choose which one you want without confusing the make-beliefs of one with the make-beliefs of the other. More complex tags allow you keep track of your imagination.

The pretense box runs into more problems here, because it's just a bag of make-beliefs, with no internal categories to know when to apply what make-belief to what story or future plan. So if a single box of imaginings doesn't explain how we do it, how does it work?

MICROTHEORIES

Let's start by broadening our focus: even in the realm of real beliefs, many, perhaps most, beliefs are only contextually true. Take some general fact about the real world that seems universally true—water is wet, blue things look blue. With a bit of thought, we can think of contexts in which these facts are not true at all. Ice and water vapor are not wet. Blue things don't look blue in darkness, or under red-colored lights. A single molecule of water is neither blue nor wet. So what is the status of your "belief" that water is wet? It's not simply believed, end of story: it's believed in certain contexts and not in others.

This suggests that beliefs also need to be contextualized in our heads. Things are true in some contexts, but not others. It's the same kind of problem we get with understanding different fictional worlds. The "many stories" problem is actually a problem with understanding reality, too.

This was discovered by an artificial intelligence project that was (and still is) trying to encode all of the commonsense beliefs of a typical person. It's called "Cyc" (pronounced "psych," like the "cyc" in "encyclopedia.") The creators, Doug Lenat and Ramanathan Guha, found that just about all of the beliefs they put in were only true in certain contexts, so they made little theories to tag the contexts in which they were true—they called them "microtheories."[56] In the microtheory of land animals, mammals have hair (with some exceptions). These microtheories can be nested—we can have a large "physics" microtheory, and within it some kind of folk "fluid dynamics" theory to help us understand what happens when we knock over a glass of water.

Although Cyc's microtheory is not intended to be a model of human thought, philosopher Jeanette Bicknell and I used it to come up with a theory of how we keep track of contextual beliefs and make-beliefs we have in different fictional stories and hypothetical situations.

First of all, remember what we learned from source monitoring: the first time we try to determine whether a memory is real or imagined, we infer from the context, using things like vividness. But once we have made that inference, we store it for use in the future. We remember that we already concluded that, say, we actually met the magician David Copperfield, rather than just having imagined it. This is the "tag" I refer to.

So for actual beliefs, they get tagged according to the contexts in which they are true. So water is dry in the context of ice and subfreezing temperatures, for example. These tags can be hierarchical, too: things are true in the context of the surface of Earth, and on Earth, there are microtheories about what water is like, how physical objects drop to the ground, and so on.

This can be applied to make-beliefs, too. Julie might have a Lord of the Rings microtheory, and maybe nested ones for each book, or the movie versions, or even different subplots in them. She has a separate microtheory for the little vampire story, which she quickly concludes should be nested in a larger microtheory of vampire lore.

The idea that "vampires exist" is not true in any real-world microtheory, but is true in the vampire lore microtheory and, in turn, the microtheories nested in it, such as the one for the little vampire story or the *Frost Moon* urban fantasy novel. In this way, Julie doesn't get imaginings in the novel confused with reality, nor with other fictional worlds.

So tagging our beliefs and make-beliefs with hierarchically organized tags help us keep track of fictional and actual beliefs.

Another interesting thing our minds can do is represent the beliefs of other people. People disagree on things about the world, and for the most part, we don't have any trouble keeping track of who believes what. For example, if you eat all of the peanut butter in the house while your kids are out playing, you'll know that they still believe there's peanut butter around. To know this, though, the representation that peanut butter is still in the house has to be somewhere in your memory. It's kind of a "belief" that isn't believed. *You* believe there is no peanut butter, though, so the two "beliefs," both that there is and is not peanut butter, both have to exist in your mind—one for you to personally endorse, and the other that you don't, but know that others endorse. How do we keep track of who believes what?

Well, it could be that something like the microtheory tagging system I've described can allow us to do that, too. You might have different tags for what Bob Loblaw believes, tags for what Bob Frapples believes, and possibly even tags for what you used to believe, but no longer do.

Similarly, we need to keep track of the beliefs of fictional people! If we go back to that little vampire story, David doesn't believe in vampires, and his belief, in that story, is false. Julie knows that in this particular story, vampires exist, but she also knows that David doesn't believe in them, and that he is wrong. She knows he's wrong, even though Julie agrees with David that vampires don't exist in the real world! The beliefs in David's mind could be a microtheory of its own, nested within the little vampire story microtheory. I'm bringing this up to show how complex our understanding of fiction can be, with layers of belief and disbelief in different contexts.

That's basically how I think we keep track of our beliefs, and belief-like ideas, be they imaginary, hypothetical, or about parts of the real world.

But the details of how we use them are more complicated, and we don't really know how it all works. For example, we might keep the facts about *Dracula* quarantined from our beliefs about the vampire movie *Fright Night*, but eventually we generate a tag of something like "vampire lore" and details from those two stories move into the larger scope of lore. When and how does this selective transfer of ideas happen? We don't know.

We also have no idea how some categorical microtheory like "vampire lore" would come to be. It must be made up of a bunch of vampire stories, but here we have a case of transfer of imaginings from a nested story to a larger, encompassing microtheory, which is the opposite direction of transfer that usually happens—that is, usually we assume that the subcategories have characteristics of the larger categories, rather than vice-versa.

Another issue is that sometimes a part of our mind believes one thing, and another part of our mind believes something else entirely. We all know what it's like to be torn about something—in these cases, it is often a case of different parts of our mind having different beliefs and values. You don't want another piece of chocolate cake, but, of course, yes you do. Part of your mind knows what's good for you, but another part of you just really wants that delicious fat and sugar. So do you believe you want it or not? When you force yourself to resist eating another piece, who is the "you" and who is the "yourself?" I'm asking rhetorically, but it's likely that "you" are your frontal lobes and conscious processing, and "yourself" is the set of old brain processes. Once I was talking to someone who pointed to her head and said, "It's all up here." But because she was a neuroscientist, I wasn't sure if she was referring to her whole brain or just her frontal lobe.

Optical illusions are also good examples of believing two contradictory things at once. You can look at one and "know" it's an illusion, but that doesn't make the illusion go away. Your early visual systems still register the illusion even though you know better. One way to look at this is that the early visual system believes the illusion, while your higher-level processes do not. Admittedly, our microtheory theory doesn't really have anything to say about this.[57]

Another problem is that we often learn real-world things, or think we do anyway, from fiction. How much of what you think you know about how courtrooms are run or how the police operate comes from movies, books, and television? For most readers, probably, well . . . all of it. People also believe things about science from reading Michael Crichton books and about history from reading James Michener. In all of these examples, the transfer is going in the opposite direction from what is proposed: fictions and imaginings become real beliefs.

How and when this happens is not understood, but it's interesting to note that people generally are pretty good at it. That is, they know what elements of fiction need to stay fiction. We read a Michael Crichton book and think we've learned something about how airplanes are built, or that dinosaur DNA is sometimes preserved in amber, but we don't believe there is a real Jurassic Park somewhere in the world, and we might be unsure whether Isla Nublar is a real island or not (spoiler alert: the island isn't real).

MEMORY ERRORS—AND ADAPTATION

When we remember something that didn't happen, psychologists often talk about them as memory errors. But to talk about a memory "error" assumes that the function of memory is recalling things that are true. But if the function is actually something different, it puts the whole idea of a "memory error" in a different light.[58]

So what good is memory, anyway?

According to Daniel Schacter, one of the very *functions* of episodic memory is to imagine the future. One of the reasons trees never bothered to evolve a memory system is because there's nothing they could really do with that information. The only "decisions" trees have to make are about directions of growth, when to drop leaves, and things like that. But these decisions can be made using perception of the current environment, and they have no need to remember anything, so they didn't evolve the infrastructure to store information.

Indeed, the plants that do need a bit of memory actually have some. The Venus flytrap has to spend a good deal of energy to open and close its trap. As such, it's not good for it to waste energy closing when there isn't a bug in there. If it had only a simple detector that shut the mouth whenever it was touched by anything, it would close every time an animal brushed by it, or when a pine needle fell on it, and so on. When bugs crawl across the hairs on the trap, it trips up one and then another. So the Venus flytrap has a short-term memory of hairs being triggered. If one hair is triggered within about twenty seconds of some other hair being triggered, only then does the plant close its trap.[59] But as you can see, this is a *very* limited form of memory. A species will have a memory system based on what it needs to survive and reproduce—evolution doesn't give living things lots of capabilities they can't effectively use, because effective use is the only way things are able to evolve.

The second thing that's required for memory to be useful is a reliable, predictable environment. If the world you're currently in does not resemble the world you've *been* living in, then it doesn't help you at all to remember what the world has been like so far. Clearly, we don't live in such a chaotic world that memory would be useless. Memory is valuable because the world is predictable. On Earth, things fall when you drop them, living things heal, but dead things don't, and playing "Girls Just Want to Have Fun" makes women scream and run onto the dance floor.

So it makes sense that a species would evolve to perceive and remember those patterns that would be likely to be relevant for guiding future behavior—and to ignore everything else. Let's take something that seems to be *irrelevant* to human behavior: the exact shape of clouds. I'm not talking about storm clouds versus puffy white ones or major cloud differences. I'm talking about the *specifics* of the outlines of puffy clouds that we can see from the surface of the Earth—a bump here, a shadow there. We tend not to remember these details because it's useless to do so (unless we notice that they resemble something that *is* relevant for life on land like a horse or something). A day later, that cloud will be gone, never to exist again. So what good would it be to be able to recognize the same cloud in the future?

So even though we have memory and trees don't, even our memories are tuned to focus on important things. We're better at remembering people we see than the exact shape of clouds or the patterns of raindrops on windowpanes. James Nairne ran an experiment testing this with word memory. He had people try to remember lists of words, and people were better at remembering words that are relevant to reproduction and survival.[60]

As we have seen, our perceptions have "errors" that can be identified using optical illusions. But scientists have also found that these optical illusions are taking advantage of shortcuts that have good evolutionary reasons for being there—they helped us navigate and survive, at least in our ancestral environment. For example, we tend to see things in terms of how we might interact with them. This is a useful thing. Everybody acknowledges that the ultimate function of memory is to help you make better decisions in the future. We assume that a correctly functioning memory would not have errors in it.

But what if these "errors" are features and not bugs? Could misremembering be beneficial? Schacter suggests the reason we have episodic memories is so that we can effectively imagine what might happen in the future. In this light, it might be that the "errors" of memory actually help make future imaginings more useful than they would if they were completely true.

How might this work? Let's suppose you have a memory of Jill as being polite. Being polite is different from being kind. But when you think about Jill, you might misremember Jill as being kind when in fact you only observed her being polite. You didn't actually observe kindness, so it's a memory error. Right?

But in the real world, polite people tend to be kind. So Jill *actually is* kind, even though you didn't observe that. So you have a memory error in that you remember her being kind even though you didn't observe it, but at the same time you are correct in that she actually is kind. The memory error actually led you to a correct conclusion. So you treat her a certain way, you expect her to be kind, she is, and you benefit from a memory error!

It might be that memory errors are helpful in this way and possibly in others. To my knowledge, however, scientists have yet to show evidence that memory errors make for better future imagining.

If this is the right way to think about memory, then why do most people, psychologists included, tend to think of the function of memory as being for veridical recall? That is, remembering things exactly as they happened.

Perhaps it is because human beings are, in many ways, general-purpose thinkers. Our minds are used for so much, we actually don't know for sure what details might be important later on and which ones will be useless. So we evolved to remember a lot of things that *might* prove useful someday, though most of it turns out to be irrelevant. There's even a board game dedicated to the pursuit of this trivial information.

Memory is necessary for being able to imagine the future and might be one of the main reasons we have memory at all. So how do people imagine the future?

3

Imagining the Future

For a long time scientists thought that only human beings were capable of recalling episodes in the past and imagining the future, but recent work by Nicola Clayton showed that, of all creatures, a bird called the "scrub-jay" appears to be able to keep specific memories of where they hide their food, and offer suggestive evidence that they consider what the future is going to be like as well.[1] But despite this slight avian competition, human beings are still the unrivaled experts at future thinking, and we do it a lot—about fifty-nine times per day. That's once every sixteen minutes.[2]

When we imagine events in the past, our imagination is using episodic memories (what happened to us), combined with knowledge from semantic memory (what we know about the world). Indeed, it's hard to imagine where else the content of imagination could come from. We also use these memories to create imaginings of possible *future* events.[3]

One reason scientists think that imagining the past and the future use similar mental machinery comes from evidence that when your memory starts to go, so does your ability to effectively think about the future. When someone gets Alzheimer's disease, amnesic syndrome, mild cognitive impairment, depression, schizophrenia, autism, or PTSD, or sometimes simply gets older, they have a harder time remembering the past, as well as imagining the future.[4] This is true, in part, because many of the same brain areas are active during these tasks. These regions are known as the "default network" and are also active in daydreaming.[5]

MIS-IMAGINING THE FUTURE

Interestingly, people make systematic, predictable errors when imaging the future. Knowing what these errors are can help you not to be misled by your own imagination.

One problem is that when we think about what might happen in the future, we rely on whatever our memory dredges up for us. We don't systematically review every possibility of what might happen. It's more like you ask your mind "Okay, I'm going to visit Italy. What can I expect?" and your unconscious mind just hands you a file with some ideas. And that file's not that thick. It might have some pictures of spaghetti and a beach.

We can think of this "file" as a narrowing of the imagination. It happens when people are asked to imagine, say, living in California versus living in Wisconsin. The most salient difference between these states, for most people, is the weather. That is, people think the weather in California is great and, to put it mildly, less great in Wisconsin. So people imagine that they would be much happier living in California, because the weather is the only thing they're thinking about. But it turns out that weather actually contributes very little to your happiness. But because people are thinking about only weather, its importance is magnified in their minds, and they make a bad prediction.[6]

We get grand ideas about whether things are good or bad—we might think of going to the movies as fun and going to the dentist as a bad experience. Of course, there are things about going to the movies that are bad, and things about going the dentist that are good. My dentist has great art on the walls and she gives me a little goody bag with a toothbrush, floss, and a piece of chocolate. When you go to the movies, you sometimes have to wait in line, deal with people talking—and, of course, the movie might stink. But when we imagine doing these things, we tend to focus only on the primary emotion we associate with them, as though *everything* about going to the movies will be great and everything about going to the dentist will be bad.

Let's take another example of something that people generally think would be really good—winning a great deal of money in a lottery. When we think of what our life would be like if we won the lottery, we tend to picture buying things, having no stress about paying bills, and having the freedom to buy a little more. What we don't think about are all the parts of our lives that would remain exactly the same—your morning routine, peanut butter sundaes still having two thousand calories, what books you like to read, the fact that there aren't enough Star Wars movies, and so on. Some of the good things in your life might even change for the worse. You might take on a more expensive lifestyle and then start worrying about affording that (this is getting on the "hedonic treadmill"). The good friends you have now might not fit in with your new, expensive lifestyle either out of envy or because they can't afford the travel you would get to enjoy. Family and friendship bonds might be strained when everybody starts asking you for money.

So although people generally believe that they would be much happier if they won the lottery, studies of actual lottery winners show that this just isn't the case.[7] Sure, they're really thrilled for a few months, but then they tend return to *whatever level of happiness they were at before.*

Well, what about something terrible happening to you? What about getting injured in a way that renders you paraplegic? Having no use of your legs sounds pretty bad, right? Most people think that if they were confined to a wheelchair to get around for the rest of their lives they'd be pretty miserable because of it.

Again, no. The same things happens. After a few years of being appropriately down about it, many acclimate and return to whatever level of happiness they were at before.[8] Note that in both of these cases there seems to be some set point of happiness that people return to. Can this be right?

In our culture, we have this idea that our happiness is a product of how good or bad our lives are. But the truth is, our situation accounts for much, much less of how happy we are than we think. Most of the differences between people (about 60 percent to 80 percent) is determined by our genes.[9]

Huh?

That's right, some people are just naturally happy, and others aren't. But you can probably see evidence of this in the people you know. We all know people whose lives on the surface are the envy of others, yet they are still moping around and complaining all the time—or are even suicidal. We also know people who seem to be unreasonably cheerful even when their lives are falling apart.

But we consistently neglect our inner happiness (or lack thereof) and expect that changing our circumstances will drastically affect our happiness. This bias affects our imaginations about the future, and what we think will make us happy.

VIVID RECOLLECTIONS

Another problem we have with both imagining the future and remembering is that vivid and emotionally salient things are most easily brought to mind. Think about what kinds of things we notice in the day-to-day world. Do you notice the pictures on your walls every day? Not likely, because you've become so habituated to them that you don't pay attention to them very often. Our minds are wired to notice changes and unusual things, so we end up in the strange situation that we remember unlikely things more than likely things.[10]

This presents problems because we have a general bias toward negativity. So when we think about the future, we remember horrible, emotional things we already know about from the past—kidnappings, murders, etc., and we end up thinking that the future will be very dark indeed. This is why people tend to think the world is more violent and dangerous than it really is. News media focuses on emotionally salient, mostly negative information, and so when we think about how the world works and what the future will bring, danger gets overrepresented in our imagination. We tend to think the world is circling the drain of the cosmic toilet, when in fact it's getting better in almost every way.[11]

But screwing up our future prediction can also get us into trouble when we're simply trying to plan our own lives to increase our happiness. For example, most people think that they need some variety in their lives to keep it interesting. If you ask people to choose ahead of time what snack they will eat, once a week for a few weeks, they will pick a variety of things. For example, suppose Liz likes bananas and peanut butter the most, but also likes chocolate milk and raisins. She might say to herself "Okay, the first week I'll have bananas and peanut butter, and the next week I'll have chocolate milk, and the week after that I'll have raisins." However, she'd have been happier if she'd just chosen to have bananas and peanut butter every week.[12]

Why is this?

Key to this finding is that the snacks are separated by a week. If you eat the same snack twice a day, things might be different, because variety becomes more valuable when the instances of snacking are closer together in time. You might have noticed this with restaurants—if you go infrequently enough, you might find yourself ordering the same dish every time you go. I think the most delicious thing on Earth is an ice cream dish that's served at Friendly's, a chain in the northeastern United States. It's called the "Reese's Pieces Sundae." I have a picture of one hanging on the wall in my house. It's got huge scoops of vanilla ice cream, hot fudge, a sweet peanut butter sauce, whipped cream, a cherry, a handful of Reese's Pieces on top, and 1,330 calories. I have not lived in a town with a Friendly's since 1989, so when I get a chance to go, I always order that sundae. They offer a three-scoop and a five-scoop version. I always get the five-scoop version, because I know it will be months before I get another chance to eat one. But one day I just wasn't that hungry, so I ordered the three-scoop version, you know, for moderation. The server gave me a huge sundae. I said, "It's not a problem, but I actually ordered the three-scoop version." She said "Oh, that is the three scoop. We increased the size of our scoops." I ate the whole thing anyway.

But if I ever find myself living in a town with a Friendly's, I might go more often, and eventually crave something else and order something like the banana split. But probably not. . . . The lesson here is that we need variety, or time in between, but not both. Yet people forget this when planning what they're going to do.[13]

PLANNING BIAS

Another mistake people often make when imagining the future is that they are overly optimistic about how long it will take to complete some task. They plan as though there are more than twenty-four hours in the day. This happens because our imaginations of the future tend to be limited to exactly what led us to generate the imagining in the first place: the task itself, with no complications. As such, we don't think about the inevitable things that get in the way of actually getting anything done in real life.[14] For example, if you were to think about how long it would take you to vacuum the house, you might use your imagination to come up with some length of time. You'd imagine it going very smoothly, with no problems or interruptions. But you'd be unlikely to imagine that you would get an important phone call in the middle of it, or that the vacuum might need to be emptied, or that you might get sidetracked cleaning up clutter or making a snack.

But we all know that real life gets in the way of our plans all the time, don't we? How do we know that? Because when we *remember* actually going through with plans, we *do* remember all of the complications. So our memories of the past have more details, but our imaginings of the future have fewer. Indeed, it even feels weird to suggest that the next time you vacuum your floor you'll get a text message from your friend with some misinformation about vitamin C that you can't resist spending ten minutes debunking. With such a bewildering space of possibilities of the kinds of things that could possibly interrupt you while you're vacuuming a floor, how, when you think about the future, are you supposed to pick some?

So even though it might be that each instance of vacuuming in our memory reveals that something always gets in the way that delays completion of the activity, from needing to go to the bathroom to receiving news of a relative dying, people show a disregard to this information when planning for the future. It's no wonder that we overcommit and overbook our schedules!

SPECIFICITY

The most vivid, detailed memories are recent ones. The further back in time the imagined event is, the less detail and vividness is experienced in your imagination. Imagining things happening in the future results in less detail, but what's amazing is that the further into the future you imagine some event to be, the less contextual and sensory details will be included, according to an experiment by Arnaud D'Argembeau.[15]

When we imagine things in the future, we tend to think of them prototypically. For example, imagine that next month you'll be at a birthday party for a friend of a friend. What kind of cake will be there? Probably you pictured a "prototypical" birthday cake. When you remember a past birthday party, though, you might remember things that deviate from the prototype, such as it being an ice-cream cake, or having lemon icing, and so on.[16]

So you might be thinking, okay, well, if we're imagining a birthday party in the past, we might be remembering a specific one, so of course it would have details that some hypothetical future birthday party wouldn't. But we imagine past things with more detail than future things even when both are completely fabricated. For example, asking people to imagine and describe a car accident that *will happen* has less detail than when you ask people to imagine and describe a car accident that *has already happened*.[17] This happens even when, in both of these cases, the car accident is completely hypothetical!

In these examples, we're thinking about generic events in the past or future, but a lot of the time we think about our own personal futures and what we'll be like then. Sometimes this future self feels very much like a different person altogether—maybe one we don't even care very much about. Do you feel like the you ten years from now is a differnet individual? How do you feel about that person? People differ in their answers to this. Some people think about and care about their future selves very much; other people not so much. It turns out that how close you feel to your future self is predictive of how well you financially plan for the future—that is, if you feel distant from your future self, you are more likely to mistreat him or her. We screw our future selves over for the benefit of our present selves in many ways: we eat food that is bad for us now and let the future self deal with the health problems and weight gain, or we spend money now, leaving our future selves poorer. I should also talk about procrastination, but I think I'll get to that later.

Several studies suggest that we think about our future selves (and, in some cases, our past selves, too) as different people altogether. We think of them more abstractly. Interestingly, imagining yourself further into the future can also change the imagined point of view from first person to third person. Emily Pronin and Lee Ross ran a study on university-aged students and asked them to imagine themselves eating meals in childhood, yesterday, right now, tomorrow, and sometime over the age of forty. The further away in time the imagined meal was supposed to take place, the more likely the students were to imagine the scene from an observer's perspective—that is, in their mind's eye, they saw themselves as though they were watching a video of themselves. For the imaginings closer to the present time, they were more likely to imagine the scene in the first person. So, for example, when someone imagined taking a bite, the food would get larger in her mind's eye until it disappeared from view, into the imagined mouth.[18]

Your connection to your future self is also related to the language you speak. There are strong-future languages that make a clear, often necessary distinction between present and future, like English, Italian, and Korean. Weak-future languages, such as Mandarin, Finnish, and Estonian, draw little contrast. People in weak-future languages save more for retirement, are less likely to smoke, practice safer sex, and exercise more.[19]

The idea that people think of their future selves as different people is also revealed in brain imaging studies. If someone feels really close to their future self, then the part of their brain that gets particularly active when thinking about oneself is more active when thinking about themselves in the future, too. But this part is less active for people who feel more distant from their future selves, and their brain activity more resembles the activation patterns we see when imagining a stranger.[20]

Our decisions about what to do in the future can depend on different things when it's the near or far future. Nira Liberman ran a study that found that when people are thinking about taking on a project in the near future, they tend to think about how difficult the task is, but when thinking about the far future, they think too much about how desirable achieving the goal would be. This results in yet another example of screwing over your future self. When choosing a distant future assignment, students were relatively more interested in how engaging the topic was, but when choosing near future assignments, they avoided the difficult ones and ignored how interesting the topic was. In another experiment, people were asked if they'd like to go to a boring lecture across the hall or to an interesting lecture across town. If the lectures were tomorrow, they chose the convenient, dull lecture. But if the choice was about which lecture they were going to go to a year from now, they chose the good lecture across town. People know they're busy tomorrow, but we always think we'll have more free time in the future.[21] If you have ever found yourself wondering "Why did I agree to do this?," then you've experienced this effect. The old you didn't worry about the future you's convenience.

Some of you might remember the Netflix queue. You could add movies you wanted to watch to it. Later, Netflix would suggest those movies to you. But they found that these movies rarely got clicked. What was happening? People would put challenging, difficult films in their queue. Stuff they really should watch, but, you know, not tonight. It became a list of aspirations. People want to watch a depressing documentary in the future, but want to watch *Borat* right now. I don't want to watch that depressing, important documentary, but my future self really should!

Netflix got rid of the queue and now suggests movies using an artificial intelligence program that picks movies based on what you and others like you have liked in the past. The result is that people spend more time on Netflix watching movies.[22]

Think about the sights of the city you live in. There are probably some that you haven't seen, but people who visit you have. You're not alone. Visitors to cities tend to do more sightseeing in a few weeks than people who have lived there for years! Without a deadline, we tend to think that we'll have more time in the future to get out and do it, so we procrastinate. It's the planning bias again. Suzanne Shu, who ran the study, also found that people who were moving away were trying to sightsee while they were packing because, in all the years they'd lived there, they'd failed to get around to the tourist sights of their own city![23]

When things in the future are hard to imagine, in a concrete, sensory way, it makes it hard to plan for them. Imagine having depression versus diabetes. The images brought to mind with diabetes are much more concrete and vivid: thoughts of dialysis, injections, and so on. The image of a diabetic injecting themselves with insulin is easier to picture, with associations of pain. In contrast, depression doesn't *look* like much at all. It's a very internal disease. As such, it's harder to imagine vividly. The difference in how clear the imagining is results in people claiming to be willing to pay more to avoid diabetes than depression, even though depression results in a lower quality of life. Depression can be devastating, but it's hard to picture, so we underestimate its effects.[24]

Studies with teenagers by Laurence Steinberg found that those who were more future-oriented were more likely to be able to wait to get more money in the future rather than taking a lesser amount of money in the present. Future-oriented people are also more concerned about environmental causes, which tend to have effects that are only felt in the future.[25]

How can we learn to be nicer to our future selves? Your connection to your future self (called "future self continuity") makes you do things to take better care of him or her. Imagination can help.

A study by Hal Ersner-Hershfield found that when people were shown images of themselves that had been digitally altered so that they looked older, they were more likely to allocate money to retirement savings in a laboratory exercise.[26] That's looking at a picture. But what about using your imagination? My colleague Eve-Marie Blouin-Hudon had people merely imagine their future selves (in sessions, over the course of weeks) and she found that they felt more of a connection to their future selves and also didn't procrastinate as much (compared to another group of people doing meditation).[27]

Another suggestive finding, this one by Albert Bandura, found that people who generate a mental image of their future-self performing some action will tend to regulate their current behavior to make it easier to do that action in the future.[28]

GOAL VISUALIZATION

Your imagination can also mess you up when you're thinking about your personal goals. Some people recommend that you visualize achieving your goals. The idea is that a clear notion of what you want will help you achieve it. Some versions of the "law of attraction," as suggested in the popular but unscientific book *The Secret,* propose this: that what we believe in our mind and think about, through mental images and self-affirmation, can attract it to our lives. The universe manifests it for you. Imagine having a lot of money, they say, and the money will come to you.

This is a very popular theory, and there's a version of it that's right. You can affect your own mind by deliberately putting it into different mental states. When you meditate, for example, you can calm yourself down. When you imagine something horrible, you might get yourself upset. As I will discuss later in the book, you can also use imagery for pain reduction and some other medical benefits. In short, thinking about this or that can affect your mind, and your mind can affect your body. Generating positive feelings in yourself, such as gratitude, hope, and forgiveness, can cause increased immune system functioning, lower stress, cardiovascular benefits, reduced risk of stroke, and less depression.[29] Scientifically, this is not controversial.

But can you imagine money into your life? Here's where it doesn't work so well. There doesn't seem to be any evidence that there's a direct relationship between the representations we entertain in our heads and effects on the outside world. Those representations can make you move your body, and, of course, you can affect the physical world with your body—making a phone call, opening a jar of peanut butter, and so on. But merely thinking about the peanut butter jar being open, and thinking that you deserve an open jar of peanut butter, will not cause the universe to conspire events to eventuate a jar to magically show up in your life and open itself. So for things outside yourself, imagining them to be true does not cause them to be true.

Sometimes the law of attraction advice is actually worse than useless. There is evidence to suggest that imagining your goal being fulfilled actually makes it *less* likely to come true.

Let's look at what happens here. Imagining that you have achieved some goal is engaging in fantasy, and one of the characteristics of fantasizing is that it feels good—that's why we do it. Your imaginings feel so real to you that you actually get a bit of the satisfaction you should feel for *actually* achieving the goal, but you get it just from *imagining* achieving the goal. Why do fantasies feel good? Because it *feels* like something good already happened to you, and that's satisfying.

The unfortunate result of this is that you are *less* likely to do the actions necessary to actually achieve these goals you imagined already having been achieved. Shocking as it seems, imagining your goal being achieved decreases your chances of obtaining it. Shelly Taylor found evidence of this in students: those who were asked to imagine doing well on an exam studied less for it and got lower grades! Similar findings found that fantasizing about the future predicted lower effort and lower success in several fields, including weight loss, romantic relationships, grades, and getting a job.[30]

So I'm trying to finish writing this book. When I fantasize about it being done, seeing it in the bookstore, having strangers come up to me and tell me how much they liked it, getting a good review, the joke I plan to tell when people ask me which of my books is my favorite, and so on, it feels good. I end up with a smile on my face. Also, when I tell people I'm writing a book, and what it's about, I get rewarded from them. "That's great!" they say, and I feel pride. I'm being rewarded without having finished it. (This might be specific to writing and other arts, but if you're writing a novel, I've read advice that encourages you to keep the story to yourself—wanting to tell people the story motivates you to write it in book form, and telling people the story allows you to express it without writing it. You should have to write to get the story out!) In support of this idea, Peter Gollwitzer ran a study in which people had goals that were important to their identity. If other people knew about the goal, then the goal seekers engaged in less action toward it than if those goals were kept secret.[31] This is, presumably, because the social approval sapped their enthusiasm for doing the hard work to achieve the goal.

Writing a book is a long, hard process. So what should I be thinking about instead? What should my self-affirmations be? I should be imagining getting up and working on the book every morning. I should imagine working on it even when I don't feel like it, and when I'm stuck, and when I think the book is terrible. I should tell myself that I have what it takes to keep at it through obstacles and difficulty. Evidence shows that visualizing the steps you'd have to do to achieve the goal helps. In Shelley Taylor's study, the students who imagined doing well on an exam got worse grades than those who imagined *studying* for the exam.[32]

Imagining the steps and the processes you need to undertake to achieve a goal activates some different brain areas than those activated when you imagine the goal being achieved. Kathy Gerlach found that imagining goal achievement made people rate the goal higher in desirability and importance. It activated the part of the brain called the "default network" (I'll talk more about this in the chapter on mind-wandering), as well as the reward parts of the brain. In contrast, thinking about the steps you need to do to achieve a goal activates the default network, the control areas, and the emotional areas, to a lesser extent. Imagining goal achievement allows people to experience the positive emotions they'd get too early.[33]

The lesson is to imagine what you should be doing to achieve your goals, rather than fantasizing about them being achieved and hoping the universe will do the work for you. That's a "secret" actually worth knowing.

Fantasizing about the future feels good. But even when just generically thinking about their futures, people tend to be very positive. In a study by Ian Newby-Clark, participants were asked to imagine events that were likely to happen in their lives. Their future imaginings were universally positive, and when asked to imagine negative things happening to them in the future, it took them longer to do it.[34]

But a problem we have with imagining the future, whether it's something as simple as vacuuming the house or as complicated as making more money, is that we overestimate the length and intensity of future emotional states, good or bad. In other words, you're not going to feel as bad or good in the future as you imagine you will. So if you imagined how you would feel getting a colonoscopy (where they push a tube into your anus to take photos of your colon), the reality is probably not going to be as bad as you imagine it will be.

In general, people think that traumatic things that might happen to them in the future will be much more depressing than they actually end up being. In fact, when most people suffer something terrible, like the death of a loved one, they recover rather well. Some even claim that their lives were ultimately enriched by the experience. This is not to discount their grief or sadness at their loved one's passing, but rather that they were able to cope and evolve psychologically and emotionally. At the same time, that trip to Barbados you've been fantasizing about probably won't feel quite as great as you imagine it will, either.[35]

Business researcher Eugene Caruso's experiments found that people have this "impact bias" for future and past events. In other words, thinking about episodes in the future makes people experience more intense emotion than when thinking about episodes in the past, even if those episodes are completely made up.[36]

Because our moral reasoning is so closely tied to our emotions, our stronger emotional feelings about the future result in stronger moral judgments about things in the future, too. Caruso found that good deeds imagined as happening in the future are judged as more good than those in the past, and bad deeds in the future were judged more harshly.[37] For example, let's take a good act, such as donating 5 percent of your income to a really effective charity like the Against Malaria Foundation.[38] When people imagine doing that, they will rate the moral goodness of the act as greater if they imagine doing it next week than if they imagine doing it last week.

Predicting how you're going to feel in the future is called "affective forecasting." There are other interesting ways we get it wrong. Suppose I asked you to predict how you'd feel if you read about a deadly wildfire that killed a bunch of people. Do you think you'd be more upset if I told you that five people died, or ten thousand people died?

If you're like most people, you'd predict feeling worse if you heard about more people dying. Why? Some death is bad; more death is worse. What's surprising is that you'd be wrong—people tend to be equally sad for both versions of the story.

Elizabeth Dunn had one group of people predict how sad they'd be if they read a story about ten thousand people dying in a fire, and another group predict how sad they'd be if they read a story about five people dying in a fire. The first group's average *predicted* sadness level was higher than the second group's. Then she brought in more people and each one either read the five people or ten thousand people dead versions of the stories, and after had them report on how sad they *actually were*. That is, the latter two groups didn't *predict* how sad they'd be, they just read the articles, got sad, and reported and reported how sad they felt. A startling thing happened: the people who read about ten thousand deaths were *no sadder* than those who read about five deaths. Our emotions, it seems, are blind to numbers.

Even if we can all agree, intellectually, that more death is worse than less death, there is evidence that small numbers are actually *more* emotionally charged than large ones. People are much more willing to help individuals than to help problems affecting lots of people. In a study by Deborah Small, people were given $5 and the opportunity to donate however much of it they wanted to a charity called "Save the Children." Some were informed about a single, identifiable victim, and others were given statistical information about the problem of starving children in a particular country. People who learned about one kid gave over twice as much as the people given statistical information.

Statistics are incredibly important. As a scientist, I know that without statistics, it's impossible to know so many things in this world. You'd hope that statistics, even if they don't work as well as individual stories, at least don't hurt. Turns out they do. In Small's third condition, people were given information about the individual child, *as well as* statistical information about the country. This group also donated far less than those who only heard about the child.[39] As a scientist who's a fan of statistics, I thought this would make me really sad, but it wasn't as bad as I was expecting.

Lots of people say, "the death of one man is a tragedy, the death of millions is a statistic." Many people think Stalin said it, but he probably didn't. Probably lots of people said it. Still, it's more compelling to think of one person saying it than lots. Kind of appropriate.

Students read Anne Frank's diary because hearing the story of one girl is more impact than hearing a bunch of numbers.

The reason we get these strange effects seems to be tied to the somewhat separate reasoning versus emotional parts of your brain. Thinking that hearing about massive deaths making you sadder *makes sense.* It makes sense to your rational mind, but your rational mind only has tenuous ideas about how your mind actually works (which you already know, because you're reading a book about psychology rather than just getting the same information by reflecting on how your own mind works). People have theories about how their own emotions work, but these theories are often wrong.

So how can we make people care about massive amounts of suffering and death? Numbers are processed rationally, but pictures, like with imagination, are more closely tied to the emotions. Elizabeth Dunn reasoned that you might be able to get a bigger emotional response by showing people pictures of the dead, rather than just telling them how many people died. Some looked at fifteen pictures of people who died, and some looked at 500. As predicted, looking at more pictures of the dead made people sadder. For experimental control, she also had other groups simply read about fifteen or 500 people dying—without the pictures, the number of deaths didn't affect the emotions at all.[40]

This is promising, but it seems likely that this won't scale up very far. For example, it seems unlikely that people would be sadder looking at ninety thousand pictures of dead people versus nine thousand. People are pattern detectors, and when you see lots of similar things, you stop seeing them as individual instances and notice the pattern instead. This is what happens with repeating visual patterns, such as wallpaper and floor tiles.

Having evolved in groups of about 150 people, our minds are not well-suited for gut instincts about the fate of thousands of people. And as such, we can't trust our feelings about them, and should use our rational processes instead. Our compassion and sympathy are generated by the old brain, and the old brain simply doesn't deal with numbers. It responds very well, though, to sensory stimuli, which is why Dunn got the results she did with the pictures. Although I have not seen a study do this, I think that even imagery cannot adequately communicate *really* large numbers of people. Think of the millions that died in Stalin's Russia—one can try to turn this into a concrete image, looking at picture after picture, so that eventually one's mind becomes numb to it.

Our minds are excellent detectors of change, but once a pattern is understood, the individual pieces that make up that pattern fade from consciousness. We get "burned out" when we hear about the same social issues again and again, and they fail to move us. So although imagination is better than statistics for making us feel compassion, the limited nature of sensory imagery, and its inability to really render the scope of huge problems, can leave us feeling flat. It seems there is no way to really make people feel more for two million people than one million (though I'd be thrilled to see a study prove me wrong).

EFFECTIVE ALTRUISM

This should be sobering information for people who really want to make this world a better place. When people want to help the world, they often judge how effective they're being by how good they feel about what they're doing. But it's easy to see how our feelings can give us a false sense of effectiveness. For example, suppose Pat meets someone who's going to die if they don't get $5,000 for some emergency treatment, and so gives the money. Across town, Chris saves two people's lives by giving $5,000—but Chris never actually met those people. It's clear that Chris did twice as much good as Pat did, but it's also painfully clear that Pat would probably feel a lot better about the donation than Chris. We're wired to feel strong emotions about things right in front of us, visceral stuff. But when problems are distant, or abstract, as are the solutions our money or time invested can generate, our emotions often won't be aroused.

But people tend to donate time or money to charitable causes based on how the charity pulls at their heartstrings. This can result in people donating an estimated $7,500 to pay some actor dressed up as batman to drive a kid with cancer around in a Batmobile.[41] Fulfilling a kid's dying wish positively drips with emotion. We can get teary thinking about this poor kid getting his dream come true. It feels better in our hearts than spending the same amount of money on hundreds of malaria bed nets that will save many lives.

But this is exactly the situation we're in. In our modern era, charities can be evaluated and rated on how effective they are. GiveWell.org reviews studies of charities and ranks them in terms of their effectiveness. With some effort, you can overcome your emotions to help you do the most good you can in this world. For an extended discussion of the effective altruism movement, I highly recommend William MacAskill's book *The Most Good You Can Do.*

It's clear that your imagination of the future has strange effects on your emotions. How else can the imagination affect our feelings?

4

Imagination, Feelings, and Morality

OUR POINT OF VIEW

Some memories are painful, and some are happy. But you're not completely locked into how you feel about them. You can actually change your feelings about them by reimagining them in different ways.

One of the unexpected effects of reimagining is whether you picture the scene in a first-person or third-person point of view. Computer game players are likely very familiar with this concept—in a first-person point of view, it's as though you're looking out of your own eyes. You can't see your head. A third-person point of view, on the other hand, pictures you in the scene as though you were watching yourself on a video. Think back to eating breakfast this morning. You can picture the peanut butter toast getting closer and finally disappearing in the bottom of your mind's eye as the toast enters your mouth. That's first person. You can also picture the very same scene in the third person, with yourself pictured in front of you, like a character in a movie. The fact that we can do this, even though it's not at all how we experienced it, is evidence of how memory recall is reconstructive. We can switch between these perspectives in our imaginations at will.

What's surprising is that the point of view you use affects your feelings about the imagining. The first-person point of view makes people focus on sensory qualities, such as the feel of the fork in the hand, the brightness of the sun, and so on. It is more in the moment and focused on your response in the situation.

The third person is a visual stepping back, but it's also a metaphorical stepping back—people picturing scenes in this way are more reflective on the *meaning* of what they're doing, and how it fits into their lives in general. For example, if you ask people to imagine going through a door, the people you ask to use first-person perspective say things like "I can feel the knob. I am turning it." Whereas people asked to use the third-person perspective say things like "I'm trying to get out."

The first-person perspective is also more emotional. When people imagine emotional events from their past, they tend to use the first-person perspective. In fact, you can reduce the emotional punch of a memory by deliberately switching to third-person when recalling it. So if you have experienced a traumatic event, such as being mugged, you might find yourself reliving the event in your imagination and also re-living the negative emotions you associate with it. But if you force these imaginings to be in the third person, they won't be so frightening.[1]

MORALITY

Imagination is also important for figuring out what is going on in the heads of other people.[2] It is a facet of our ability to empathize, but the two can often be at odds. We misunderstand other people a lot. One of the ways we get it wrong is that we reliably underestimate how nice other people are going to be. We like to think of ourselves as good people, but think of everybody else as more selfish.

This is shown in one popular experimental setup called the "prisoner's dilemma." In it, you and another person can independently choose to cooperate or "defect" (screw each other over). If you both cooperate, it's best for both of you. But if the other person defects, it's also better for you to have defected. One study by business researcher Eugene Caruso found that people will typically cooperate about 60 percent of the time. However, this percentage changes when they are asked to empathize with the other person involved with the dilemma. But contrary to what you might expect, because people think of themselves as more good than others, after considering the thoughts of others, the probability of cooperation goes down to 27 percent![3]

Chillingly, empathy does not always help.[4]

It's well known that our emotions are an important part of our moral feelings. You might feel moral outrage, or jealousy, for example. Because imagination can generate emotions, it can also affect our feelings of moral judgment. A moral judgment is when you experience or hear about some scenario in which somebody does something, and then judge whether the person was behaving ethically or not.

In experiments, these scenarios use the imagination. People construct imagined situations in their heads based on what they are told, and then they think about them to make some kind of moral decision. For example, philosophers Shaun Nichols and Joshua Knobe gave people the following scenario:

"Imagine a universe (Universe A) in which everything that happens is completely caused by whatever happened before it. This is true from the very beginning of the universe, so what happened in the beginning of the universe caused what happened next, and so on, right up until the present. For example, one day John decided to have french fries at lunch. Like everything else, this decision was completely caused by what happened before it. So if everything in this universe was exactly the same up until John made his decision, then it had to happen that John would decide to have french fries."

Fewer than 5 percent of people thought that people were morally responsible for their actions in Universe A. When we think about acting morally, or immorally, we think about free will and making choices. For many people, a deterministic universe as that described in the text makes moral choices seem kind of impossible. So, they reason, people in Universe A have about as much moral responsibility as a lawn mower does. The example in the text above involves eating french fries. What happens when we make the example a bit more visceral?

"In Universe A, a man named Bill has become attracted to his secretary, and he decides that the only way to be with her is to kill his wife and three children. He knows that it is impossible to escape from his house in the event of a fire. Before he leaves on a business trip, he sets up a device in his basement that burns down the house and kills his family."

This gets you in the gut a bit more than the french fries story, doesn't it? Indeed, when given this scenario, a full 72 percent of people said that Bill is morally responsible for his actions, even in this universe where everything is caused deterministically from what came before. Just changing the action in the story affects people's idea of whether there is free will or not, and that affects their moral judgment, to the tune of a whopping 67 percentage points.[5]

Let's look at another example: suppose a runaway train is about to kill five people. In this situation, is it okay to push an innocent person into the path of a runaway train, killing that one person, to save the five that would have died otherwise? Think about what your opinion on this is before you read on.

There are two answers to this question. The first is a rights-based approach, called "deontology," which favors the rights of the individual being pushed. According to deontology, you may not violate the rights of the pushed person, even if it would save more lives. The other view is a more public-good view, called "utilitarianism," which says that the act is good because it saves five lives rather than one. Different people give different answers to this question, and many find it difficult to decide what's right and wrong.

One can think about this problem rather abstractly, say, in words, or one can try to vividly picture the scenario. Elinor Amit looked at a bunch of people and ran tests on them to see if they were more visual or more verbal thinkers (visualizers versus verbalizers). She found that their intellectual style predicted their answer to this moral dilemma: the visualizers were more likely to think of it as wrong (the deontological stance), and the verbalizers were more likely to think it right (the utilitarian stance).

Asking people what was in their "mind's eye" when they were deciding revealed that people were more likely to imagine killing the person pushed, rather than imagining the five lives saved. That is, when imagining, they tended to image the harmful means rather than the beneficial ends.

So it looks like people who naturally visualize had more rights-based judgments. If the act of visualizing was causing this shift, then, she reasoned, if you interfere with their ability to visualize, the effect should be reduced. She did exactly this. In the experiment, she had them do a challenging visual task *while* making the moral decisions about pushing someone in front of the train. The task was this: they saw a series of shapes on a computer screen, and for each, shape they were asked if the shape they were looking at was the same or different from the shape they saw *two shapes ago*. This interferes with imagination and visual imagery, because it requires holding several shapes in mind at once. This made it much harder to vividly picture the runaway train situation because it kept the visual areas of their brains pretty busy while they were supposed to be making moral judgments. As predicted, the visual interference made people more utilitarian. That is, when she interfered with their ability to visualize, they favored the "greater good" moral stance, and thought it was more acceptable to push the person in front of the train.[6]

So it seems pretty clear that visualizing a moral situation makes you think more about rights than the greater good. We also see, from the french fries versus murdering your family example, morals are triggered more by visceral scenarios. But why would this be? To make sense of this, recall the distinction between the old, emotional brain and the new, more analytical brain.

The emotional and visual systems are all a part of the old brain. The old brain has more of an animal-based, intuitive morality. This means that it has evolved to have a kind of morally appropriate to respond to situations one is likely to face in a preindustrial society. And one of the hallmarks of preindustrial societies is that it's difficult to physically hurt someone from a great distance.

In one version of the train problem I described above (called "trolley problems" in the literature), two similar situations are compared. In both, a train is headed to kill five people, and you are judging whether to choose to have one person die instead. But in one, you're pushing somebody in front of the train, and in the other, you're pulling a switch to redirect the train to another track, where it will kill only the one person instead. So in both, the idea is that you're sacrificing one person to save five. But people tend to give different answers for these two problems. In general, more people are willing to *pull a switch* to kill one person to save five than they are to *push somebody* to have the same effect.

One of the important differences between these situations is that pushing somebody means actively putting your hands on them. This is a very visceral thing to do and relates more to the kind of situations our ancient ancestors might have faced. It's kind of like pushing somebody over a cliff or onto some sharp rocks. Because our ancestors faced similar ones, they evolved to feel that these situations are morally wrong.

But there's nothing in the ancient world like pulling a switch to do anything at all, let alone kill somebody. As a result, for many people, pulling a switch doesn't feel as bad, even if you (or more specifically, your new brain) "knows" that the effect is ultimately the same—sacrificing one innocent person who would otherwise live. This is one of the reasons offered for why people differ in their moral responses to these hypothetical imaginings, but we don't know for sure of what's going on in the brain quite yet.[7]

So, the theory goes, because imagination is closely tied to evolved morality, engaging in visual imagery activates more of those old circuits, resulting in an older, evolved, rights-based moral judgment, and a relative deactivation of the rule-based, cool reasoning that the new brain specializes in. If you think about some moral situation visually, you're more likely to think about it using evolved morality, as opposed to a more principled, reason-based morality.[8]

Those are some examples of how imagining a moral situation can affect your judgment about it. But can you use your imagination to make yourself a better person? Judging others is one thing, but what about how you actually act in the world? If you imagine doing good things, will that encourage you to actually do good things?

The answer seems to be yes. Brendan Gaesser found that when he had people imagine helping people in need, they were more likely to intend to help people in real life. And the more vivid the imagining was, the stronger the desire to help people.[9] This seems to jibe with the fact that morality is closely tied to emotion and feelings about morality. Similarly, Arber Tasimi found that children who thought about their own good past behavior (but, interestingly, not others') acted more generously.[10]

It sounds like it makes sense, but it also seems to contradict some other findings. Recall that imagining the steps one takes to achieve a goal (such as studying for an exam) helped people study, but imagining the goal being fulfilled (such as getting a good grade) did not. In fact, it hindered it.[11] Imagining your goal being completed gives you a great feeling that can, ironically, sap your motivation to achieve that goal in real life. Gaesser had people imagining helping others, which presumably involved imagining both the steps involved, as well as the positive outcome of helping. So if they were imagining the outcome, why didn't this sap their desire to actually do it like it did in the studying for exams experiment?

Another reason to think it wouldn't work is because of something called the "moral credential effect," which is how your feeling like you are a good person allows you to license yourself to do bad things—or to not do good things. Sonya Sachdeva found that people who wrote self-congratulatory essays chose to donate only about a fifth as much money to charity as those who wrote self-critical essays. Many experiments involving subjects, such as racial prejudice, diet and health practices, and energy consumption, have shown similar "licensing" effects.[12] And, as we see in many studies, imagining doing something is treated the same way by much of your mind, as actually doing it. So when you imagine doing good, we have two reasons to think that it might *inhibit* your tendency to do good in real life, not the other way around.

These studies come from different literatures and seem to come up with opposite conclusions. Moral licensing predicts compensatory behavior, making you act badly, and work from the "moral identity" literature suggests that it makes you more good. What the hell, psychology?

I think Paul Conway might have figured out the answer to this paradox. In his study, he showed that thinking about good things you've done in a more specific, concrete way (rather than an abstract one), or thinking about good things you've done recently, causes a compensatory behavior. That is, it makes you less likely to do good deeds. But if people think about being good in an abstract way, or about good things done in the distant past, it reinforces a moral identity and results in people being *more* likely to act in a good way.[13] So does thinking about being good make you more good? Well, it depends on *how* you think about being good! If you think about it in the abstract, or about the steps involved with doing good, or about things you've done that were good in the distant past, or about being good as part of your identity, it might help you be a better person. But if you imagine being good vividly, perhaps using mental imagery, or think about good things you've done recently, or focus too much on how great it would feel to do good, you are likely to think you're already "good enough" and choose not to do as many good things.[14]

A study by Gert Cornelissen also sheds light on this problem. When asked to imagine having done something good, people with a more outcome-based mind-set were less likely to be good in the future, and people with a rule-based mind-set were more likely to be good.[15] These mind-sets roughly correspond to utilitarian and deontological ethical theories.

So thinking about being good can, if done correctly, encourage you to be a better person. There's also evidence that you can use your imagination to become less prejudiced. Irene Blair found that you can change certain stereotypes simply by imagining people of a certain gender or race, doing or being something that is counter to your stereotype. Across five experiments it reduced multiple kinds of stereotypes.[16]

So, used with care, imagination can make you act better. Can it make you feel better? Can you use your imagination to become kinder, less prejudiced, and just happier? It certainly seems like thinking about sad things makes us sadder, and thinking about happy things makes us happier. What does the science say?

HAPPINESS

Thinking about the bad things in your life can be harmful. Len Lecci showed that pondering the regrets of your life encourages reduced life satisfaction and discourages effective coping with life events.[17] But what happens when you think of how your life could be worse than it currently is?

When you imagine how the world might be, rather than how it is, it's called "counterfactual thinking," because it's counter to the facts. Counterfactual thinking is one aspect of imagination that can influence our happiness and general satisfaction with life. Keith Markman showed that thinking about how your life could be better tends to make you feel worse, with increased negative feelings, such as regret, remorse, and disappointment. The quality of something is often judged in comparison with some other specific thing. So if you think about how much better your life *could* be, you are likely to have a dimmer view of your actual life. Just as a candy will taste less sweet to you after eating gummy worms all day, imagining what your life could be like affects what you think of your life as it is.

But wait a minute—isn't imagining how your life could be better a form of fantasizing? And doesn't fantasizing feel good? I mentioned earlier that imagining achieving your goals can sap your motivation to do the steps you need to do to actually achieve them. This is because it feels good to imagine your goal being achieved, and that reward kind of makes you feel like you've already achieved it. But why does this work, given Markman's finding that imagining your life better tends to make you sadder, not happier?

Fantasizing feels good, but there's a dark side to it. Gabriele Oettingen found that fantasizing about a happy future leads to a short-term boost in happiness, but more depressive symptoms down the road.[18]

So, if thinking about how great your life could be makes you feel worse, can thinking about how *bad* your life could be make you feel better? Yes. Keith Markman also showed in his experiments that if you think about how your life could be worse, it increases positive feelings, such as surprise, joy, and relief in avoiding disaster.[19] It sounds simple, right? When thinking about a bad life, yours looks good, and vice versa.

But it's not that simple.

One time, when I was a child, my sister and I were sitting in the back seat while my mother was driving. My mother was trying to pull out onto a road. Visibility wasn't good, and at one point she considered moving forward. She didn't, though, and a good thing, too. A big truck zoomed right where she would have been. If she had pulled out, we might all have been in a terrible car accident. But since she didn't, if she then thought about this past that almost was, it would have been like thinking about how her life could have been worse and should have made her really happy. Right?

She wasn't. She was rattled and upset and kept thinking about how she almost killed herself and her children. Counterfactual thoughts like this often lead to mixed emotions: there are positive emotions resulting from feeling fortunate that the bad thing didn't happen to you, paired with anxiety—*that could have been me.*[20]

Why does this happen?

Suppose a golfer is thinking about Tiger Woods. Mr. Woods is a great golfer, and comparing herself to him might make her feel bad about her own ability. But if she was a member of a minority group, the idea of Tiger Woods might be inspiring. By considering herself in the same class as Mr. Woods, she could share in the joy of his success. If she *contrasts* herself with Mr. Woods, it could make her feel bad. If she *identifies* with Mr. Woods, she won't feel bad. She might feel good.

The reverse happens with thinking about how your life could be worse. If, in your imagination, you put yourself in the situation, then you might feel anxiety or some other negative emotion. But if you focus on the contrast between your life and what you've imagined, it can feel good. One of the reasons the car example above is compelling is that my mother might feel both of these contradictory emotions at once, caused by different processes in her mind trying to make sense of them.

What all this means is that you can use your imagination to improve your mood, but you have to be strategic about it. Thinking about how your life could be worse can help you have gratitude about your life as it is. For example, I love being a university professor, and sometimes I think about what my life might be like if I had another job. *Any* other job. This makes me happy in that it fosters appreciation. But if I think about *losing* my professor job in the future, or about someone else who lost their job whom I can identify with, it freaks me out and causes anxiety.

Reflecting on what you're grateful for can provide even more benefits than downward counterfactuals. Gratitude is a complex phenomenon involving both beliefs, attention to memories, and emotion. You can think of it as the perception that one has received some benefit because of chance, of the good intentions of another (rather than some benefit earned, deserved, or worked for).

One way to exercise your gratitude muscle is to keep a gratitude journal, writing regularly about what you're grateful for. A study by Robert Emmons compared people who kept a gratitude journal to people who journaled about things that were wrong with their life, and even to people who recorded how their lives were better than the lives of the less fortunate. Relative to these other groups, the gratitude group reported being more alert, enthusiastic, determined, attentive, and to have higher energy. They were also more likely to help someone else or provide emotional support. They experienced physical benefits, too, such as longer and better sleep, and engaged in more exercise. Friends and family of people who practice gratitude report that they seem happier, more pleasant, outgoing, trustworthy, and optimistic—it's also better for your health.[21]

What does gratitude have to do with imagination? Well, deliberately recalling episodes from your life, and evaluating them in terms of gratitude is a kind of imagination, but because people vary so much in how vivid their imagery is, some are using imagery and many are not. Although studies have not distinguished imagery versus nonimagery gratitude interventions, because we have seen how imagery is so closely related to the emotional parts of the brain, it would not surprise me if deliberately engaging in imagery while doing gratitude practice enhanced its effects.

Gratitude is thinking about your life as it is, but we often use our imagination to think about what might have been. These are things we know to be false, but temporarily suppose to be true. It is a sandbox where we can build an imagined reality and see what happens. It's a simple imagining of a possible world, a world that might be, or might have been.[22]

When do we engage in counterfactual thinking? Typically, we do it when we want to change or feel the need to understand the world. For example, we often will do it for negative or exceptional events, but less so for things that are routine. People will spontaneously think of counterfactuals when things *almost* happen. For example, people think of more counterfactuals when they miss an airplane by five minutes than when they miss a plane by an hour.[23] If the world treats you well, or you missed something good by a long shot, why question it?

We often create imagined worlds simply to help us understand what people say. For example, when you read a novel or hear a story, your imagination is active and is, indeed, necessary for comprehension. Sometimes we can get so drawn into these stories, recreated in our own imagination, that we stop attending to our actual surroundings. For example, in a particularly exciting part of a book, it might not even enter your mind for several minutes that you're sitting in a chair, at home, or even reading at all. Psychologists call this "transportation," because you feel transported to the world created by the author—or at least, the version of it you are creating in your own mind.

Psychologists have devised clever experiments to figure out what makes a story compelling, what kinds of people are more likely to experience transportation, and why people seek it out. At first blush, it might seem an impossible phenomenon to study scientifically. Although measuring transportation is not as straightforward as, say, measuring the mass of a rock, Melanie Green and Timothy Brock devised a questionnaire that reliably measures people's tendency to get lost in a good book. It includes statements that respondents are asked to agree or disagree with, such as "I could easily picture the events taking place."[24]

Transportation is closely linked to the concept of flow, pioneered by creativity researcher Mihaly Csikszentmihalyi, which is the good feeling you get when you are completely absorbed in a task and lose track of time, forgetting about other concerns in your life for a while. People can experience flow when creating art, engaging in sports, or numerous other activities. Interestingly, reading creates the flow experience more than any other activity.[25]

Why do people seek out transportation experiences? One of the reasons is mood control. When feeling bad, sometimes people will read a happy or funny story to put them in a better mood,[26] and indeed, literary transportation is more appealing to people who are feeling bad about themselves. Perhaps because they are using transportation as a kind of escape, people in the laboratory made to feel bad spent more subsequent time watching television, presumably to make themselves feel better.[27]

People also get better transportation effects if they can relate to the story. For example, it's easier to experience transportation if the season in the story matches the actual season, or if the social environment is one you're familiar with, such as a fraternity or an arts organization.[28]

Many parts of our minds believe what is happening in the story is actually real, and transportation can make these stories seem even more real. What ends up happening is that stories that we know full well are complete fictions can end up changing beliefs we have about the real world. One study, by Elizabeth Marsh, found that avid readers of romance are less likely to use condoms, presumably because they are influenced by the "swept away by romance" trope in the literature they read.[29]

Of course, transportation has beneficial effects, too. Steven Pinker suggests that the introduction of literature to the masses is one of the driving factors in reducing violence worldwide—because fiction, perhaps more than any other art form, has the ability to bring the reader intimately into the head of another, which encourages empathy with people who are different from you.[30] A lifetime of reading correlates with having social skills, such as perceiving what others are feeling and thinking.[31] One striking study found that fiction promoted compassion even more so than real-life experiences. In an experiment by Phyllis Katz, one group of white children worked on a shared task with some African American children, while another group read a story with African American fictional characters. The students who read the story had more positive attitudes toward African Americans than the students who actually interacted with them.[32]

Writing a novel, and even just reading one, is a profound act of imagination and perhaps a key to a better future.

I think I'll take a break from writing this and work on my novel. . . .

5

Hallucination

Hallucinations might seem to be an extreme form of imagination. We think of hallucinations as being a psychiatric or neurological disorder (as did some of the ancient Greeks), but in the Middle Ages, and even in some parts of the world today, they are thought to be caused by supernatural agents, such as angels or demons.[1] Even in the mid-1800s, scholars debated whether or not hallucinations could be considered normal, whether they formed a continuum with sensations, dreaming, and mental imagery, and the relationship between hallucinations and other mental states, such as dreams or ecstatic trances.[2] These issues are still discussed by scholars today.

The popular contemporary view of hallucination being a symptom of a mental disorder isn't completely accurate. Lots of people have them, even without mental illness. Personally, I've had several auditory hallucinations, always when I'm falling asleep. I might hear my name, or a doorbell, but then it turns out to be nothing. I also sometimes see moving spots of light in the periphery of my vision after I blow my nose. Since I got a cell phone, I occasionally feel phantom vibrations in my pocket.

The situation is further complicated by cultural factors. Cultures of the Upper Amazon use a hallucinogenic drug, ayahuasca, as a spiritual guide. The Siona people believe that the experiences they have on the drug show them an alternate reality. The Schuar people believe that the drug experiences *are* the reality, and what we see day to day is the hallucination.[3] Although we can't prove that they're wrong, for this book I'm going to treat normal waking experience as that of the real world, more or less, and hallucinations as fictions generated by our minds.

HALLUCINATIONS AND GRIEF

Bereaved people who are suffering the recent loss of a loved one can hallucinate that that person is still around. A survey by W. D. Rees found that almost half of bereft people surveyed experienced hallucinations of their dead spouse—some involving full conversations.[4] Thirty to sixty percent of widowed individuals have hallucinations of their dead spouse, and about 50 percent of people have hallucinations when they are close to death—even if they have never hallucinated before. They are often very emotional, personal, and have accompanying feelings of tenderness and warmth.[5] Twenty percent of children have imaginary companions, but because they are not believed to be real by the children who host them, they are missing what many consider a key feature of hallucination.[6]

So just because someone has hallucinations doesn't necessarily mean they are mentally ill. That said, hallucination is a symptom of several mental illnesses.

Hallucinating can sometimes be described as mistaking imagination for perception. Like imagery, hallucination is when one has a perceptual or sensory experience that was not caused by the thing in the world that normally cause the experience—usually seeing or hearing things that are not there. Many of the same areas of the brain are active when perceiving and hallucinating. One explanation for hallucinations is that victims suffer from dysfunctional source monitoring. That is, victims don't correctly identify the source of their own experience. They think it's perception when it's actually imagination.[7] Are all hallucinations simply imaginings that you mistake for reality? This is a very old idea, going back at least to 1768.[8]

To begin with, we need to differentiate between what is a hallucination and what is a product of imagination. My definition of hallucination sounds a lot like the definition of imagination in general, but there are some key differences. One is that imagination is often (though not always) voluntary. Though imagination can sometimes come unbidden, as with PTSD, or when you can't help but picture something disgusting as someone describes it, in general, we can turn our imagination on and off just by deciding to do it. We also tend to have more control of our imaginations, whereas the content of hallucinations is often surprising, and we have no conscious control over it.

Another important difference is that we don't tend to believe in what our imagination produces. For example, if you imagine Elton John standing in front of you, you probably won't mistake him for the real thing and ask him why it's taking him so long to make a version of "Candle in the Wind" for Anna Nicole Smith. You know it's make-believe. Many hallucinations are believed—though, not all, as we will see. Being able to tell that your mental image or hallucination isn't real is called "insight."

But sometimes even our voluntary imaginings can be mistaken for reality, or, in other words, imagination can lead to hallucination. In the 1960s, Theodore Barber asked people to imagine hearing the song "White Christmas." Five percent of those people reported that they thought a record of the song was actually being played![9]

Like imagination, hallucinations can have three basic forms, in terms of how they relate to what you're perceiving in the world at the time. The simplest is when the hallucination seems to be on the inside of one's eyelids, or permanently in the same part of the visual field, no matter where you look. You might hallucinate moving sparks on the periphery of your vision, and looking left and right doesn't allow you to focus on them—they're always on the periphery, no matter where you look.[10] If you glance at the sun (please don't), you will get an afterimage for a little while. This is a very minor hallucination that follows every movement of your eyes. With these kinds of hallucinations, when your eyes are closed, the hallucination (or image) might appear to be vaguely in front of you, but is not tied to the external environment at all. This is probably what you do when asked to imagine a green number seven, for example. It's not imagined to be out there in the environment, but just sort of floating in front of you no matter where you look. This is kind of like the psychological version of your character's status display (such as the health bar) in a 3-D shooter game, which stays on the same part of the screen no matter which way you turn.

Getting more complex, we have imagination and hallucination that is projected into the world, as when you are looking at a wall and imagining a picture on it. This is kind of like the augmented reality of imagination and hallucination. You might be watching a jazz band and imagine an extra band member who plays a solo on an air horn. When you look away from the stage, she's not in your visual imagery anymore, and when you look back, there she is again.

Some hallucinations are completely immersive, where you perceive yourself to be in a totally different environment, with little or no perceptual influence from your eyes. This can happen to people when they're dreaming, under the influence of the drug DMT, or during temporal lobe seizures. This is like the virtual reality version of your imagination. In normal imagination, this happens when you imagine being somewhere different and when you're dreaming.

We think of hallucinations as being sensory in character—seeing or hearing something that is not there. Delusions, in contrast, are more typically about belief and meaning. Someone might have a delusion that they are someone famous, for example, or have some "understanding" that the United States is controlled by aliens. We might think of delusions as nonsensory hallucinations—hallucinations of belief, rather than of sensory experience. Both delusional and hallucinatory thinking might be thought of as using the same "jumping to conclusions" thinking style, with delusion applying to belief and hallucination applying to perceptual experience.[11]

Interestingly, in terms of their experience and in the brain areas that get activated, hallucinations seem to be more closely related to perception than to imagination.[12]

PHANTOM LIMBS AND MYSTERIOUS MEN

We can also have hallucinations or delusions (and also, often, imagery) in "senses" that we don't often think about. You can, for example, sense your body's position. When you wake in the morning, before you move or open your eyes, you can tell where your arms and legs are. How do you do this? It's not because you're looking at them, and it's not because of your sense of touch. You just have a feeling for where they are, and this is a form of sensation called "proprioception."

Sometimes when people lose a limb they are left with the feeling that it's still there—a phenomenon called "phantom limb." Sometimes when people are immobilized for a long period of time they might have proprioceptive hallucinations, such as feelings that they have more or fewer legs, or that they are particularly long or short.[13] This is one of many examples of how sensory deprivation can cause hallucinations.

Another "sense" we have is whether someone is near us. When someone is, say, in the room with us, we have a sense of where they are. Sometimes we are wrong, and the person has left the room and we are surprised to turn and find she is no longer there. Other times we get the feeling that someone is there (such as the feeling of being watched) when there was never anybody at all.

This feeling of presence can also be hallucinated, sometimes with Parkinson's patients on a drug called "L-Dopa." It might feel like someone is in the room with you, even when there is nobody there. This also happens to people without Parkinson's disease when they are on long wilderness journeys with lots of social isolation. Explorer Sir Ernest Shackleton reported that he and the two other people with him felt the presence of another, incorporeal person on the journey with them in Antarctica.[14] Whether or not these are proper hallucinations might be a matter of semantics. The feeling of presence is not, strictly speaking, sensory. You don't see or hear anybody, you just have a feeling that they are there.[15] This might involve spatial imagery, which is related to multiple senses.

When we see something we know well, we get a "feeling" of familiarity. This effect feels so natural that it's a tough thing to notice until something goes wrong with it, as it does in déjà vu. Neurologist Oliver Sacks describes a patient who would get attacks several times a day, which made completely new things feel familiar. The patient would even do strange, completely novel combinations of motions to prove to himself that the feeling was only an illusion. That is, he would move his body in a way he *knew* he'd never done before, and then notice that it still felt familiar. This is how he would get insight.

Similarly, people can fail to get a sense of familiarity in environments they know well, such as their own home. Patients *know* that they are at home, but it still doesn't feel familiar. The fact that a person can be in this state is evidence that the *knowledge* of something being familiar is separate from the *feeling*.[16] With direct brain stimulation, you can elicit a general feeling of familiarity without even having anything in particular to feel familiar about![17] Some hallucinations involve a strong feeling of understanding, or of intellectual acuity—that one, perhaps, "understands everything," or could solve any problem.[18] As we can see, there are many psychological states that are not sensory, per se, but can be the subject of certain kinds of hallucination.

CAPGRAS, SCHIZOPHRENIA, AND OTHER CAUSES OF HALLUCINATION

Another hard-to-classify example is a common symptom of Capgras syndrome, which is often caused by excessive consumption of alcohol: that loved ones are not recognized as such, but instead are regarded as perfectly similar duplicates. For example, the wife of a Capgras patient might come into the hospital room, and the patient will accuse her of being a phony and ask where his real wife is. This seems to be caused by a disconnect with the emotional response that the patient is used to getting in the presence of someone he cares about. When he looks at his wife, whom he knows he loves, but feels no familiarity, he concludes (this is all subconscious) that it must not be his wife at all. The only available conclusion is that it must be a cunning duplicate. Capgras not only impairs the victim's sense of familiarity, but also their sense of evaluation, which, if intact, would allow them to have some insight into their condition.[19] Again, the sensory experience is the same, so this might be better classified as a delusion rather than a hallucination. Neurologist Oliver Sacks reports a "duplicate" symptom that appears to not have any emotional relevance. One of his Alzheimer's patients recognized the nursing home during the day, but at night felt that she'd been transferred to a "cunning duplicate."[20]

These interesting cases aside, what we know the most about are the sensory-like hallucinations. I'll start with the most complex hallucinations and work my way down to the simpler ones.

Hallucinations are most famously a symptom of a devastating disease, schizophrenia. This is not to be confused with dissociative identity disorder, which is what used to be called "multiple personality disorder," or "split personality." Schizophrenia is a psychotic illness characterized by a break from reality, and people who suffer from it can experience delusions and hallucinations.

Though people can experience hallucinated sounds, such as ringing or even music, most hospitalized persons with schizophrenia hear hallucinatory voices, the most common kind of hallucination in neurological and psychiatric disorders. It is important to know that up to a third of people who have nothing wrong with them at all also hear voices at some point in their lives—hallucinations, in the non-clinical population, are common. So if you hear a voice in your head once in a while, it doesn't necessarily mean you have schizophrenia, though schizophrenic voice hallucinations (in Western cultures, anyway) are more often persecuting, jeering, or threatening than the voices heard in nonclinical people. When the rest of the population hears voices, including those whose hallucinations are caused by brain disease or hearing loss, their character is usually unremarkable, or incomprehensible.[21] Brain imaging suggests that the same brain mechanisms are at work for people with psychiatric disorders as for the rest of the population when they hear these voices, and we can't tell from the brain activity why voices in the head are so much more distressing in schizophrenia.[22] Hearing voices can happen in dementia and epilepsy, as well as in schizophrenia, but it can also be caused by extreme stress or isolation.[23] Because people in the Western world associate hearing voices in their head to be a sign of being "crazy," they are reluctant to talk about them, making it hard to study its prevalence in the general population. But the studies we have suggest that between 4 and 15 percent of people experience them.[24]

As you might imagine, during auditory hallucinations, the parts of the brain that process sound are more active.[25] The auditory association cortex (the superior temporal gyrus) seems to be "busy" during hallucinations and is less responsive to sounds in the environment while a hallucination is happening. This might be why listening to music or talking can decrease or stop auditory hallucinations, by causing interference.[26] Brain areas used for the generation as well as the understanding of speech are active during verbal hallucinations. These findings support the idea that the verbal hallucinations have to do with speech processing. In fact, there is subtle movement in the mouth and larynx during verbal hallucination, and keeping your mouth open can sometimes prevent the hallucinations from happening![27]

In some non-Western cultures, hearing voices in your head isn't even considered to be a sign of mental illness and can be viewed favorably. Even in the West, we have precedent of voice-hearers who were (in their time) considered generally mentally healthy, such as Joan of Arc, Galileo, and Socrates. Even today, most people who hear voices, but are not mentally ill, believe not only that they come from some external source, but that they come from benevolent spirits.[28]

About 50 percent of psychotic patients report that these hallucinated voices replaced their own moral conscience.[29] Voices in the head are sometimes regarded as a sign of being divinely chosen in many religions, including Buddhism, Pentecostal Christianity, Sufism, and shamanism. It might not be surprising that others interpret the things said by hallucinated voices as divine truth. Hallucinations are mysterious, and religion often provides explanations and interpretations of things we don't understand.

Most schizophrenic hallucinations are auditory (though, interestingly, Africans tend to have a higher frequency of visual hallucinations), but patients can have hallucinations with many senses.[30] People with schizophrenia often hallucinate full characters, complete with audio and visual experience. However, these characters are usually rather shallow, not even as deep as many characters in literary fiction.[31] This might be related to the fact that during schizophrenic hallucinations the medial prefrontal cortex is relatively deactivated (normally active during daydreaming), an area of the brain thought to process understand the self and the thoughts and feelings of others.[32]

If hallucination is experiencing something that isn't there, that something need not be a whole object—it can be an unreal characteristic of a real object. For example, sometimes people will look at one person but see five, or see some event once but experience it happening again and again. The original event was real, but the repetition of it was a hallucination.[33] Some call hallucinations that are provoked by real sensations "functional hallucinations."[34]

Another kind of hallucination involves the feeling of sensed presence, as well as a visual hallucination of another person, but the person is yourself. These are called "autoscopic hallucinations," and there are three kinds. The first is simply an autoscopic hallucination, which is when the hallucinator sees and image of him or herself, projected in the world, but their sense of identity is still with their body. The second kind is the out of body experience, which is like the first one, except that the hallucinator feels that the *real* self is in external space, and they are seeing their bodies from another disembodied position (often above). The third is heautoscopy in which the hallucinator sees a double but is unsure which is the real one![35] Many hallucinations are the result of sensory deprivation, and it turns out people who are more in touch with their bodies like dancers have fewer out-of-body experiences, suggesting that autoscopy might be related to not getting enough input to your proprioceptive system.[36]

A variety of drugs can cause hallucinations, including LSD, hashish, DMT, mescaline, Artane, psilocybin mushrooms, and morning glory seeds (the active ingredients of which resemble LSD in their chemical structure). These hallucinations are like schizophrenic ones in that they are often multisensory and soaked with meaning. One important difference, however, is that drug-induced hallucinations seem more likely to be pleasant experiences—if they weren't, people probably wouldn't take the drugs.[37]

Many effects are perceptual alterations—seeing colors change, things multiplying, or changing size—but sometimes there are delusion-like effects of feeling that one understands everything, or senses infinity, or feels that everything is interconnected. For some, a single experience like this can feel so profound that it changes one's outlook for the rest of their lives. One interesting aspect of these experiences is that while the hallucinations, as such, are not always truly believed (the drug user might know that they are drug-induced perceptual hallucinations), the delusions *are* believed. For example, one might look at one's skin and hallucinate that each individual cell is breathing, or blowing smoke rings.[38] At the same time, one might get the feeling of vast interconnectedness of things in the universe. The cells blowing smoke rings are not believed, but the associated belief about the interconnectedness of all things *is* believed.

Some drugs, such as Artane, might produce hallucinations that are believed completely, such as hallucinated people or monsters.[39] One of the things that is fascinating about hallucinations is that they take on a variety of complexities. DMT, for example, the active ingredient in ayahuasca, produces "full-world" hallucinations: you experience being in a completely different environment, full of characters, as opposed to the hallucinations integrating with your perceptual environment. Users of DMT even prefer to be in a dark room with their eyes closed so that the hallucinated world and the perceptual world don't interfere.[40] This is not true for LSD, which creates hallucinations that integrate with, or are distortions of, your perceptual world.

When we think of hallucinations we tend to think of highly complex, multisensory ones that appear to be situated in the world, such as the schizophrenic hallucinations of people as depicted in *A Beautiful Mind*, intertwined with delusions and ideas of reference. The hallucinations caused by drugs and schizophrenia are perhaps the most complex. The ones caused by epilepsy are a bit simpler.

Epilepsy is characterized by seizures, where there is a sudden electrical discharge in the brain.[41] When these occur in the higher levels of the brain (in the parietal or temporal lobes), the hallucinations can be very complex, perhaps involving animated cartoon-like figures. At even higher levels, a person might enter a dreamlike state in which they are in a completely imagined, hallucinatory, multisensory world.

But when seizures affect the lower-level perceptual parts of the brain, the results are experiences of simpler images, such as spinning balls of light.[42] These findings make sense, because seizure location predicts the kind of hallucination people have: the hallucinations will be related to the kind of perception that the affected brain area processes. So if you have a seizure in the part of your brain you use to see spots, you'll hallucinate spots.

Charles Bonnet syndrome is also characterized by hallucinations, but most people with this syndrome only hallucinate shapes, colors, or patterns, and never see full objects, people, or scenes. They might not even think to mention their symptoms to their doctors.[43]

In more complex Charles Bonnet hallucinations, the person might see faces, or some kind of notation, such as text, numbers, or musical notes. About a quarter of people with this problem have some kind of text hallucination. They might see words, always meaningless, spread across the car's dashboard, on the stairs, or on some other surface.[44]

Still more complex Charles Bonnet hallucinations might involve complex scenes such as people in costumes marching around. These are projected into space—that is, to the hallucinator it appears that the people are in the environment with them. There is never an auditory component to them—it's purely visual. And they are never interactive. They don't engage with the hallucinator at all. Further, the people and places are almost always unfamiliar. They are plausible, but do not represent specific people or places from experience. All of this suggests that Charles Bonnet syndrome is a disorder that generates hallucinations from a very specific, "low level" part of the visual system—one that has representations of colors, lines, and patterns, and at higher levels generic objects, people, and notational symbols, such as letters and numbers—devoid of the richness of meaning and emotional charge that often accompanies dreams and schizophrenic hallucinations. Indeed, "interpreting" the hallucinations, as one imagines happens in talk therapy sessions, seems to be a dead end with Charles Bonnet patients.[45]

Other Charles Bonnet hallucinations involve strange combinations of things, such as animals on people's heads or plant life coming out of people's cheeks. The fact that there are these seemingly random combinations, and because they are essentially meaningless to the hallucinator, suggest some kind of random activation of what things look like in the visual areas of the brain.

Neuroscientist William Burke found correspondences between the size of the patterns in hallucinations and the size of the active parts of the brain's visual area V1. He suggests that blobs of neural matter in these areas give rise to patterned Charles Bonnet hallucinations.[46] Although Charles Bonnet hallucinations are simpler than many schizophrenic hallucinations, they are still complex in that they are projected onto the world and can be rich in visual detail. That is, one might see musical notation on a wall, but if one looks away from the wall, one doesn't see it anymore. Turn back, and there it is. Somehow your mind is keeping track of where that hallucinated text is supposed to be.

Charles Bonnet hallucinations can be extraordinarily vivid—even more rich in color and detail than what can be seen with the eyes. This marks another important difference between hallucination and run-of-the-mill imagination, which tends to be less vivid than real life and certainly less vivid than hallucinations. Further, they tend to be exotic and elaborate—the people hallucinated might be wearing outrageous costumes, and the music notation seen is often more ornate than any score in the real world. We don't know why this happens.[47]

It seems obvious that people with vivid mental imagery would be more prone to hallucinations, or at least would have more vivid hallucinations. But, at least for people with schizophrenia, this does not seem to be the case. Study results are all over the place, but there appears to be a pattern of finding no relationship between hallucination and the vividness of one's imagery.[48]

It can be easier to have insight (that is, identify a hallucination as such) if is less complex. A cross-hatch pattern suddenly appearing when you look at a blank wall is unlikely to fool anyone, because it's so implausible. More complex hallucinations, such as seeing someone walk into the room, are more believable. It requires more critical thought to recognize it as a hallucination. But it is characteristic of Charles Bonnet syndrome that the hallucinations rarely actually fool the hallucinator, perhaps, in part, because they are usually so incongruous with the environment, the disease does not affect critical thinking, and because of their lack of meaningfulness and emotion, they almost never lead to delusions or persistent false beliefs (assuming the absence of other mental disorders, such as dementia).[49]

Some Charles Bonnet hallucinations are described as being like watching the theater, in that the performances are unpredictable and uncontrollable. This brings to mind the idea of the Cartesian theater, a concept introduced (and rejected) by philosopher Daniel Dennett. The idea of the Cartesian theater is that somewhere in your head is "you" and that "you" can watch and listen to other things going on in the mind. For example, when you look at a dog, the dog's image is represented in your brain and the subpart of your brain that is "you" looks at it. The reason Dennett (and most cognitive scientists) reject this notion is that it doesn't provide any explanatory power, because, in this case, the question of how perception works is merely shuffled from the entire brain to the part of the brain that we are calling "you." Thus, is the danger of positing the existence of a little mind inside of the big one.[50]

Though I agree with Dennett in the case of perception, Charles Bonnet syndrome makes an interesting case for the Cartesian theater in hallucination. It seems that some parts of the mind are working on their own, generating imagery that some other part of the mind observes, as though it actually were watching a stage play. The observing part of the mind can find the imagery amusing, boring, annoying, and can believe or not believe that it represents anything real. But it does appear that some functions in the mind generate imagery, and others perceive and judge it, and often can't control it. It sounds a lot like a theater to me.

HALLUCINATION AND THE VISUAL SYSTEM

It appears that different mental and brain disorders differ in where the hallucinations are happening in the visual system. Charles Bonnet syndrome hallucinations, for example, often seem to be the result of activation of the lower-level visual system, where at the other end of the spectrum, schizophrenic hallucinations, which can be indistinguishable from reality, often involve brain activation at the higher levels of the visual system (among others).

The occipital cortex processes visual information rather early, and focuses on angles and orientation, and other elements of elementary hallucinations. In visual perception, this information is fed to higher levels, such as the inferotemporal cortex, which detects more complex visual and spatial features, such as animal forms, plants, objects, letters, and faces. Complex hallucinations like complex imaginings are likely combinations of activations of these features.[51] In contrast, simple hallucinations—single colors or sounds, for instance—tend to work with more local, smaller regions of the brain.[52]

Neuroscientist Wolf Singer used a brain-imaging method to observe which parts of the brain were relatively active during imagination versus hallucination. The main difference was that imagination is accompanied by activation in what are called the "executive areas of the brain" (normally associated with the prefrontal cortex, right behind your forehead). The executive areas are active for conscious, deliberate control of what you're doing, which, in this case, reflects the voluntary aspect of typical mental imagery. The fact that this area is not active during hallucination (at least for hallucinations caused by sensory deprivation) suggests that unlike imagination, these kinds of hallucinations originate elsewhere—presumably the ventral visual pathways in the brain, which tend to get excited over time if they don't get any input from the eyes.[53]

MIGRAINES

People can also experience simple hallucinations when they are having terrible headaches known as "migraines." These hallucinations are called "migraine auras." They might start as one thing, such as zigzags, and over time change to checkerboard patterns or something else—the same as those experienced under the influence of many hallucinogenic drugs. As the electrical signal of the migraine travels across the brain, it activates different perceptual detectors that correspond to these elementary hallucinations (bodily distortion hallucinations are also common with migraines). That is, rather than the hallucinations coming from meaningful memories of the individual, they are merely activations of perceptual areas of the brain. The migraine activity activates the detector for a zigzag line, and we experience one. But simple hallucinations like these are rarely believed. That is, migraine patients rarely think the lines they see are something in the actual environment. They're pretty obviously hallucinations, though.[54]

These simpler hallucinations also occur with Charles Bonnet syndrome, parkinsonism, sensory deprivation, migraine, epilepsy, drug intoxication, and the hallucinations you get while falling asleep (hypnogogic hallucinations). They do not seem to reference the individual's desires, fears, goals, emotions, or memories. They seem utterly irrelevant to anything.[55] For these reasons, simple visual hallucinations probably don't have a lot to do with imagination, if we think of imagination as drawing from memories.

Hallucinations that originate in the higher areas of the brain have a different character altogether. Brain imaging studies have shown that hallucinations of faces were associated with activation in the superior temporal sulcus, while objects and extended scenes are associated with the ventral occipito-temporal cortex, the part of the visual system that specializes in facial recognition. These are strange and sometimes horrifying perceptual hallucinations in which the parts of peoples' faces are hallucinated as being too big, too small, or otherwise distorted into grotesque visages—huge teeth or multiple eyes.[56] In contrast, the more simple geometric hallucinations sometimes experienced with Charles Bonnet syndrome are related to the kinds of features that can be detected with early visual areas, such as V1 and V2.[57]

Post-traumatic stress flashbacks might cause someone to experience a combination of hallucination and delusion to convince them that they are at war again and that the people around them are enemies—a multisensory, emotional, convincing virtual reality generated by their brains.[58] Needless to say, these experiences are much more emotional experiences than seeing zigzag lines. Although verbal and visual hallucinations get a lot of attention, you can have hallucinations in other senses, too.

OLFACTORY HALLUCINATION

Your sense of smell, or olfaction, can also experience hallucination, particularly under conditions of deprivation. Sometimes when someone loses their sense of smell, they can start to hallucinate. Neurologist Oliver Sacks describes a patient who lost his sense of smell and was delighted to find it had returned when he smelled his morning coffee—only to find, when tested, that his smell had not returned, and he was merely hallucinating the coffee smell.

What is mysterious is that olfactory hallucinations can be so vivid, while at the same time most people have very weak olfactory mental imagery, and rarely dream with it.[59] Normally we can't imagine smells vividly, as we can visual and auditory information. For example, I have fairly vivid visual and auditory imagery: I can think of scenes from movies and experience them again in my head. But although I know what peanut butter smells like, I can't image that smell in a way that feels like I'm really smelling it. Given that we think that imagery and hallucination use a lot of the same parts of the brain, this is mysterious.

Just as people might visually hallucinate something weird, olfactory hallucinations can be weird. It's easy to understand why. Your nasal cavity has more than five hundred different kinds of receptors, and the activation of different combinations of them can result in trillions of potential olfactory experiences. As a result, many olfactory hallucinations are utterly indescribable! We couldn't possibly have enough words for all of them.[60]

Ten to twenty percent of people who lose their sense of smell end up experiencing olfactory hallucinations. This happens to be the same percentage of people who get Charles Bonnet syndrome upon losing their vision, suggesting that some similar process is causing both.[61]

We have lots of senses we don't think about very often, and these can have hallucinations, too. For example, we have lots of what are called "interoceptions," which are basically perceptions of what's going on in your body. These include feelings of hunger, sexual drive, the need to urinate or vomit, itching, the passing of time, that feeling you get when somebody scrapes a chalkboard with their fingernails, and so on. And as you might expect, many of these interoceptions have analogs in hallucination. A few times I've woken in the middle of the night with the distinct feeling that my body was many times larger than it really is. Other bodily hallucinations might make one feel that their body is too small, or that they have no body at all.[62] There are also feelings of time distortion, temperature hallucinations, and even hallucinated pain.

BELIEF AND INSIGHT

As we have seen, not all hallucinations are believed. Some, who through brain damage have either the left or the right side of their visual field impaired, will get hallucinations that they immediately recognize as such. People with Charles Bonnet syndrome will sometimes believe in their hallucinations at first, but quickly realize what's going on. Geometric migraine hallucinations are almost never believed. At the other end of the spectrum, schizophrenic hallucinations can be utterly convincing. Some people who are blind from brain damage cannot see anything, but because of their vast hallucinations claim to not be blind at all! When they run into furniture they might claim that the room is poorly lit or that the chair had been moved.[63]

Knowing that your hallucination is not real is called "insight," but it's not an all-or-nothing thing. Insight might be delayed, partial, or even fluctuating.[64] Insight is more common for hallucinations that are more readily explained by brain problems, and is rarer in hallucinations caused by what are considered more mental or psychiatric problems.[65]

Are hallucinations with insight different from ones without, in terms of how they work in the brain? So far, the research is inconclusive.[66]

Given that similar brain functions are recruited in high-level hallucinations and imagination, could imagery practice cause hallucinations to happen? Many religious traditions involve the practice of rituals that heavily involve the use of imagination.

Tanya Luhrmann, in her anthropological study of American evangelical Christians, found that many used practices of guided imagery that involved a focus on sensory detail. With practice, your imaginings can get more vivid, and you get more absorbed in them. Eventually, a person can hallucinate images of gods or other scenes of spiritual importance—even Luhrmann did. The religious interpretation is that you are practicing to have spiritually attuned consciousness. An alternative explanation is that through practice of vivid imagination, hallucinations of emotionally meaningful imagery results. Resembling what can sometimes happen with drug-induced hallucinations, a single spiritual experience can cause, or strongly reinforce, a lifetime of faith.[67]

Similar effects have been found in Amazonian shamans, who will sometimes describe relationships with hallucinated jaguars. Generally, these shamans are more practiced and experienced. Over time, their hallucinations become more precise, multisensory, and can even occur on demand.[68]

Why does this work? Imagining something makes you better at detecting it in the world—what we might call "priming," or an "imagery gain." For example, if you imagine a tone, you are more likely to hear that tone in the real world. Suppose there is a tone that is so quiet that you can barely hear it. People who imagine the tone will be better at hearing it than those who don't. This gain, though, can also result in hallucination, for the same reasons that using a camera in low-light conditions often results in little blips and artifacts (we might think of these as low-level camera hallucinations). These religious traditions, through the use of guided imagery, might be priming the perceptual system so much, turning up the "gain," that people think they're actually seeing what they imagine.[69]

Turning up the gain in this way probably makes people expect culturally appropriate things: the Amazonian shaman hallucinates a jaguar, a rural African might hallucinate about ancestors, and a Christian might hallucinate the Virgin Mary.

Even these religious people don't ignore the possibility of hallucinations due to mental disorders, or just as products of the mind. Cultures that treat some hallucinations as divine insight take pains to distinguish them from others.[70] Simon Dein interviewed members of a Pentecostal church. They told him that they asked themselves whether the voices they heard were in accordance with their understanding of scripture or came with a feeling of peace. If not, the hallucination was interpreted as nondivine. One man with bipolar disorder reported that the hallucinations caused by mental illness were "very pushy."[71] Some Christians also take pains to distinguish between voices from God versus voices from demons.

Some religious hallucinations are taken so seriously that they affect the whole course of the religion, and the course of history. Famous examples include the Buddha's hallucinations under the Bo tree, Moses's burning bush, and Arjuna seeing Krishna. A study of 488 societies worldwide revealed that 62 percent of them used hallucination as a part of ordinary ritual practice. Because hallucinations are often not reported as such (they are often classified as "revelation," or something like it), this is probably a low estimation.[72]

Moving on from religion, we see culturally appropriate hallucinations during a phenomenon known as sleep paralysis. In this disorder, people, between sleep and waking, feel as though they are awake, can't move, often feel pressure on the chest, making it difficult to breathe, feel abject terror, and often hallucinate the presence of malevolent figures in the room with them (if you've had experiences like this, you should talk to your doctor). What's interesting is that the character of the hallucinated figures matches expectations based on culture. Although sleep paralysis is something that happens in all cultures, it gets different interpretations. The old Western idea of the succubus, a demon who sits on your chest, is likely a sleep paralysis hallucination. As people hear about the idea of a demon on your chest, when they get sleep paralysis they hallucinate exactly that. In this way the belief in demons gets reinforced. But in other cultures, the monster is different. Sometimes it's the "old hag," and sometimes it's "ghost pressure." In modern Western society, much of the belief in alien abduction is likely due to sleep paralysis hallucinations.[73]

WHY WE HALLUCINATE

So why do hallucinations happen, and what is their relationship to imagination? As we have seen, there seem to be several ways hallucinations can happen.

One way seems to be what is best described as "dream intrusion," which is when part of your mind is dreaming during a waking state, and the dream contents are experienced as a hallucination. This is likely how hallucinations work in narcolepsy, Parkinson's disease, certain kinds of dementia, and in the hallucinations you can get when you're falling asleep or just waking up, like sleep paralysis.[74] We often have no insight for these, but this is temporary. If we wake completely, we often quickly realize it was all in our heads. We tend to believe our dreams while dreaming, it is thought, because the critical areas of our brains are quieted. It stands to reason that they would also be quieted during dream intrusion, but it is very difficult to get imaging data for dream intrusion hallucinations, so we just don't know yet.[75]

Too much activity in a sensory area of the brain can also cause someone to experience hallucinations. This is how the hallucinations of epilepsy and migraines are formed. This was found when parts of the brain were stimulated by experimentalists. Stimulating the visual area caused the experience of visual hallucinations, and stimulating the auditory area caused auditory hallucinations. Hallucinations that occur in only one sense (for example, visual hallucinations that have no sound component) are almost always the result of over-activity in the corresponding brain area.[76]

Too little activity can also cause hallucinations. Low-sensory experiences created in the laboratory reliably cause visual hallucinations that resemble those of Charles Bonnet syndrome. William Bexton ran a study in which he tried to cut off as much sensory stimulation as possible—people spent days in a soundproof room, with goggles that only allowed them to see vague lightness and darkness, and gloves and cardboard cuffs to reduce tactile sensation. The participants tended to fall asleep, wake up, try to entertain themselves with their own imaginations, and then finally they started hallucinating. What is interesting is that, in this experiment, the hallucinations started off simple but over time got more complex: changes from light to dark, dots, lines, then geometrical patterns. Most of the participants then experienced wallpaper-like patterns, then isolated figures and objects (with no background). Later the hallucinations appeared three-dimensional. About a third of the participants experienced hallucinations so complex that they were described as being like cartoons. In all cases, though, the hallucinations were essentially meaningless, lacking in emotion or in relevance to the participants' lives.[77]

What is interesting about this progression is that it seems to mirror the order of processing in the visual system. Light enters the eye and activates neurons on the retina, and from there on in that information is processed in successive stages of increasing complexity. First edges are detected, then higher-order shapes and other features. At the highest levels of perception, we get what we normally, consciously experience as vision. It seems that over time, sensory deprivation seeps through the visual system, and as the deactivation progresses upstream, more and more complex hallucinations result.

To understand this, recall that the act of vision, for example, involves what we call "bottom-up" activity from the eyes, and "top-down" expectation from visual memory. The combination of what we sense and what we expect determine our visual experience.

The sensory areas are usually busy doing their jobs: processing sensory input from the world. So, for example, the visual areas are busy processing what's coming in through the eyes. One idea is that this constant stream of input from the outside world *inhibits* the top-down activity in the visual brain areas: expectation and imagery from memory. This is why vision *loss* can actually cause hallucination—without the normal bottom-up input, the visual area is overactivated with top-down expectation and imagery, and starts making things up, as in Charles Bonnet syndrome.[78] Most people who hear voices in their heads report that the problem is worse when they are alone, which also suggests why occupying the language areas with real voices helps reduce the hallucinated ones.[79] If this idea is right, it's sensory input from the outside that keeps our expectations in check, and explains why sensory deprivation would also cause hallucination.

There is evidence to suggest that the reason meditators sometimes experience hallucinations is simply because meditation is a form of sensory deprivation.[80]

So it sounds like people who have more top-down processing would have more hallucinations. Sounds like a great theory, doesn't it? One thing wrong with it is that it predicts that hallucinators would be more affected by perceptual illusions, which are widely believed to be caused by top-down influences. But the opposite is true. Hallucinators are *less* susceptible to many of them.[81]

Computer models of beliefs in the perceptual system suggest that we hallucinate when there is high uncertainty, which might explain why sensory deprivation can cause hallucination—with nothing to perceive, our minds have no idea what is out there.[82]

INHIBITION PROBLEMS

Another theory of hallucinations concerns excitation and inhibition in the brain. But before I explain how that works, let's talk a bit again about perception.

When we look at a scene, we are using lots of information to make sense of it. Let's suppose that you open a jar of peanut butter, so you're expecting a brown substance inside. You look and see the color brown. Your belief that it's peanut butter in the jar is simultaneously reinforced by your expectation of peanut butter, as well as the information from your eyes.

In this example, the expectation of peanut butter is making the perception of the presence of peanut butter in your mind more likely. Your expectation primes your peanut butter detector, making it more likely to be activated. This expectation is the top-down effect.

But your perception of peanut butter is also influenced by seeing the brown color in the world, producing the bottom-up effect. Seeing brown activates all brown things in your perceptual system, making perception of them easier.

So "brown" activates "peanut butter," and then "peanut butter" activates "brown." This would be bad if it became vicious circle. If there was a positive feedback loop like this, you can imagine someone getting more and more sure that it's peanut butter just because of a feedback loop in their own mind of brown activating peanut butter activating brown activating peanut butter. . . . [83]

The way the mind deals with this is through something called "inhibition." Most neural connections in the brain are excitatory, but many are inhibitory. In perception, inhibitory connections keep information from being counted more than once. That is, the fact that you saw brown should affect the probability that you believe peanut butter is in front of you, but it shouldn't have that effect again after the belief in peanut butter affects your belief that you're seeing brown, or the system would spin out of control.

Now let's take a different example. Suppose you observe a man being yelled at. This generates an expectation that the man will be upset by this. You see his facial expression. It might be a neutral facial expression, but you are likely to interpret it as stressed out, because of your expectation. If you saw the same facial expression on someone being complimented on a job well done, you might interpret that same neutral facial expression as proud. The expectation of emotion affects your perception of a neutral face. Filmmakers take advantage of this through the Kuleshov effect in which audiences see the appropriate emotion in a neutral face based on the context—sad if it's a funeral, angry if someone they care about was hurt, and so on.[84]

Sometimes our senses violate our expectations. Suppose you are around someone with an injury, so you expect to see some blood. But near the person you see a blue liquid. We now have two sources of information for scene interpretation that don't make any sense—we expect red liquid, but we see blue. We see blue liquid, but expect blood. When the mind is functioning correctly, it should be less sure what is being seen in incongruous situations like this: is there blood or not? The blueness coming from the eyes is reducing the probability of the interpretation that "there is blood," while the knowledge that there was an injury is increasing the probability that "there is blood." These two forces inhibit each other, leading to a lack of a firm conclusion.

But if the inhibition is unbalanced, either too much in favor of either the high or low level, it can lead to a bad conclusion—more confidence in either one interpretation or the other ("it's definitely blood" or "it's definitely not blood") than is justified. You might have someone insisting that blue paint is blood, simply because their expectation got too much weighting in their brain. This imbalance of inhibition has been suggested to be a reason somebody might have a hallucination or a delusion.[85] In this case, there would be too much top-down influence.

But problems can result from there being too much bottom-up influence, too. Suppose you see a woman who is standing, waist-deep, in some dark water. You cannot see her pelvis or her legs. If you're healthy, you'll probably interpret that scene correctly—that there's a whole woman there, even though you can't see her bottom half. That's using what you know about the world to help you understand incomplete information coming from your eyes. But if there is a problem with your mind and brain such that these expectations are given less priority, and you rely more on your senses, then you might trust your senses too much, and think that the woman's top half is resting on the surface of the water, with no legs at all. This would be paying too much attention to your senses and not using your expectation and knowledge enough.

One time I was lying in bed, and out of the corner of my eye I saw a shadowy bit moving on the wall. What was it? My eyes were telling me that something was moving over there. But my knowledge about the world says that that's impossible. Nothing in the world looks and moves like that.

I also know that the human visual system has noise in it, and once in a while it will generate stuff that's not real. So I thought nothing of it, except that I might put it in a book someday. (If I'd been the kind of person who believed in supernatural creatures, I might have come away from that same sensory experience with a very different interpretation of what happened. And written a very different book.)

But if my inhibition system were messed up, and my brain was giving too much weight to my senses, I might believe that, in spite of what I thought I knew about the world, I really saw something extraordinary—some kind of creature, perhaps, that looks like a shadow, but could appear and disappear from view. When people rely too much on their senses, any sensory noise could potentially get turned into a hallucination. A patient might hallucinate seeing peanut butter, just because they saw a bit of the color brown.[86]

It could be that one way that delusions and hallucinations happen is because of problems with inhibition in the brain, where expectations get interpreted as sensory information and sensory information gets misinterpreted as a prior expectation.

A further theory is that hallucinations occur as a result of recalling memories of things we've experienced in the past, without being aware that all we're doing is only a memory retrieval. This theory has been used, primarily, to explain voices in the head. In support of it, some of the same brain regions are involved in the encoding and recall of auditory memories as well as auditory verbal hallucinations, and 49 percent of voice hearers recognize what they experience from a memory of a conversation they've had before. However, the parts of the brain used in auditory perception are also active, as we've described above, suggesting to some that it's not a memory retrieval at all, but perhaps just over-activity in the auditory part of the brain.[87] We don't know which theory is right. A better way to put it is that we don't know which *set* of theories is right, because it's possible that the brain hallucinates voices in more than one way.

Generally, lack of insight is thought to be a problem of experiencing something and then misattributing it to some external source. So, for example, one might retrieve a memory of someone saying "You should eat some peanut butter," or the language generation areas of your brain might generate the same sentence, but rather than realizing that the sentence came from one's own mind, one thinks it comes from somewhere else—perhaps an alien, perhaps a person in another room, or the CIA sending messages directly into your brain. Your ability to know how you know things, and the source of your experiences, is the "source monitoring" that I wrote about in an earlier chapter.[88]

People with schizophrenia, and people who hear voices, tend to have source monitoring problems in general. They are more likely to attribute actions and thoughts they have to others—meaning that they might have a thought, but think it's actually someone else's thought. Or their arm moves, but they think some other entity is controlling the arm. Not only do they have trouble remembering where a memory came from, but they often forget temporal information, too—when they learned it and in what context. They also tend to recall more irrelevant memories. Some researchers believe that these effects explain the misattribution: when a fragment of memory is recalled in isolation, without any of the normal contextual or temporal cues, it feels strange, and it feels like it came from nowhere. The resulting confusion leads to a belief that the thought was implanted by an outside entity.[89]

DYSFUNCTION OF THE FORWARD MODEL SYSTEM

One way source monitoring might go wrong, and, thus, lead to hallucinations, is in the movement of the body. When you move, your motor system does the moving, and the perceptual system perceives the results. What normally happens when you move your body (for any reason) is that a plan, a motor schema, is sent to the perceptual areas to give it a heads up for what's coming so it's not surprised by the movement. For example, suppose your mind (perhaps unconsciously) wills your arm to reach for a jar of peanut butter. When you do this, the motor parts of your brain send signals to your perceptual areas, telling them what to expect: "Okay, I'm getting the peanut butter now, expect to have one in the hand soon"—something like that. You won't pay as much attention to the feeling of the hand on the jar as you would if you felt the same thing *without* having moved your hand. In other words, the motor schema sent makes you pay less attention to the expected sensory results that follow. This allows your mind to focus on new, unexpected things and not be distracted by what it already knows is coming. *Of course it feels like there's a jar of peanut butter in my hand. I just reached for it!*

But if those systems are not communicating effectively, the mind gets confused. So some hallucinations are caused by people failing to appropriately anticipate what they intended to do. If there is a dysfunction in this system—either the message is delayed, corrupted, or doesn't arrive at all—then your perceptual system will be surprised by what you experience—your touch system for your hand didn't "know" it should expect a feeling there, so it's surprising: *I moved my hand, but I didn't will it to!* Because your perceptual areas experience of grabbing the peanut butter comes without any forewarning, your mind can then interpret the motion as caused by external control—by aliens, or the FBI, etc., in a disorder known as "Alien Hand Syndrome." The motor system is not surprised by the movement, but your perceptual system is.[90]

This mechanism has been suggested as the explanation for why you can't tickle yourself (personally, I have no trouble tickling myself if it's on the bottom of my feet, but in general, science shows that people can't).[91] Tickling seems to be a reaction to non-dangerous touch to vulnerable parts of the body, and even rats can be tickled.[92] But if your mind knows you're the one causing the motion, the perceptual system doesn't get alarmed and you don't get the tickle experience.

If this is true, then putting in a delay between the decision to tickle yourself and the skin actually being touched should make the feeling more ticklish, because it will seem less and less likely to your perceptual system that you're the one doing the tickling. To test whether delays can cause an attribution of external control, Sarah-Jayne Blakemore ran a clever study. What she did was set up an apparatus where somebody would move a bar, and it would cause a foam rod to stroke the hand. In general, if someone moved the bar and, thus, the rod to "tickle" their own hand, it would not feel ticklish. In the experimental condition, the decision to try to tickle oneself had a bit of a delay, so that the foam rod moved a bit after the bar was activated, so that the people in the experiment felt the sensation later and later. It turns out that the longer the delay is, the more people felt tickled by their own actions.[93] The motor schema is sent, and your perceptual system prepares to be tickled, but then the stroking doesn't show up on time. It shows up later, and the mind is less sure that you actually were initiating the stroking. We might call this a "tickle hallucination." Blakemore also ran a study showing that people who are prone to hallucinations found it easier to tickle themselves than everybody else did.[94]

This system also helps you keep track of the world when you move your eyes: the motor schema for eye movement allows your perceptual system to expect that there will be a view change. Your eyes move several times a second, but we don't really notice this. The reason is that the part of your brain that moves the eyes sends a message at the same time to your visual system, letting it know where the eye is going to move. In this way, your mind can keep track of what appears to be a stable environment in the face of lots of eye movements. If your perceptual system didn't know where the eyes were moving, then perception of the world might be extremely difficult like trying to make sense of an environment by watching the video of a camera being moved in a random, jerky way. You move your eyes down and see the floor. Your perceptual system knows that there was an eye movement down, so it knows that the floor it sees is downward from you.

But there are potentially two ways that the visual system might "know" that the eyes are moving. It might get a message from the motor system ("Hey, I'm moving the eyes down.") or, potentially, it could adjust its sense of where body parts are. In other words, you know you moved your eyes down because you felt the muscle movement. The first is paying attention to the report on the motor command, and the second is based on observing the result of the movement itself. So which one does the visual system use? The proprioception system perceives that the eyes moved, or does it just use the message from the motor system?

This puzzle was figured out in 1931 by Alois E. Kornmüller in a fascinating, weird experiment. The idea was to paralyze the eye muscles. So what would happen is that the person's motor system would *try* to move the eyes, but then the eyes wouldn't respond. When the person tried to look to the right, one of two things might happen: it might be perceived as normal, with nothing unusual happening, which would suggest that the perceptual system is paying attention to how the eyes *actually* move, rather than *assuming* they moved, based on the motor system's command. The other thing that might happen is that the room would be perceived to suddenly shift to the left—if you look right, but the image on your eye doesn't change, the only possible interpretation is that the *room* lurched around at the very moment your eyes moved!

The difficult part of conducting this experiment is that we don't know how to paralyze *only* the eyes. So they used the drug curare, which paralyzes every muscle except your heart. This includes the diaphragm, so the participants had to be hooked up to a respirator. So what happened when they looked to the right?

The room seemed to shift left! The visual system gets a message from the motor system, and then doesn't bother to check the proprioceptive system to make sure the command was executed. Such is the heroic length science had to go to in order to find out if the perceptual system used muscle movement or motor messages to help make sense of the world.[95]

So that's how we can get hallucinations and delusions regarding the movement of body parts. We generate a movement, and the perceptual system doesn't appropriately anticipate the effects. But what about imagination? When we generate something in our imagination, normally we know it's just an imagining. But what if the perceptual parts of our minds don't know that an imagining has been generated?

It's possible that some visual hallucinations also happen because of problems with the generation system miscommunicating with the perceptual system. Your mind imagines something visual, but the message from the generating part doesn't reach the perceptual part, and it feels like it wasn't imagined, but rather put in your head by some outside entity (God, the government, or some other external being). Or you think you're actually seeing it and get a hallucination without insight. This kind of hallucination is truly mistaking your imagination for reality. It's like Alien Hand Syndrome, but for imagination.

This account might explain voices in your head, too, because one thing that makes verbal hallucination different from visual hallucination is that it has a strong motor component—that is, when you're talking, your mind has to plan how it's going to move the mouth and other body parts to generate the sounds. Your mind plans to say something with inner speech, but something goes wrong with the motor schema's message to the perceptual system. As a result, the parts of your mind that understand language treat it like it didn't come from you. This is one theory of how generated inner speech can be misattributed to external sources.[96]

It also turns out that when healthy adults experience inner speech, the auditory cortex is less active. But in patients who hear voices in their heads, they get activation there. This likely gives their inner thoughts more of a physical character and can make it seem like they're coming from an external source.[97]

With actual speech, as opposed to inner speech, when a healthy person decides to say something, Broca's area (a part of the brain) plans what words to say. Then a message is sent to the motor areas for generating speech, and a copy is sent to the auditory cortex, which is in charge of listening. When the person speaks, she hears herself, and when what is heard matches the copy, the auditory cortex will take little notice of it and tag it as being self-generated. But with a dysfunction of that system, the auditory cortex might interpret the voice heard (one's own) as being someone else's or controlled by someone else.[98]

There's a saying in neuroscience: if the brain can do things five different ways, it does all ten. It appears likely that hallucinations can happen for several reasons. All problems with delusion and hallucination boil down to some kind of source misattribution, in that we mistake something coming from ourselves for something originating in the outside world. But not all of the ways people can hallucinate seem to be mistaking imagination for reality. Though all hallucinations are internally generated (if they're not, perceptual mistakes tend to be classified as "illusions," instead), different hallucinations differ in *how* the brain creates them.

I've talked about six ways, and there's a seventh. Let me summarize.

First, some hallucinations might be memory retrievals that don't get recognized as such. Similarly, we might imagine something but don't recognize it as an imagining—though, keep in mind, all memory retrievals are, in some sense, reconstructions, so these two might differ only in degree. Many verbal hallucinations seem to be like this, and sometimes visual hallucinations (during an occipital seizure, for instance) involve hallucinating a past experience as happening now.[99]

Second, they might appear because of problems with inhibition, where runaway feedback processes eventuate in a hallucination—if there are problems with the inhibitory processes, then expectations of partial perceptions can get amplified beyond what they should, resulting in a hallucination or a delusion. For example, the sensory areas are activated because of the imagination. This changes expectation with respect to perceiving new things in the environment. If there is a failure of inhibition, these expectations can be interpreted by the perceptual system as sensory information, resulting in a hallucination. This is a feedback problem in the perceptual system and doesn't seem to have much to do with imagination, if we think of imagination as using memories to think of possible situations.

Third, you can have a problem with your perception communicating via motor schemas—a dysfunction of the forward modeling system, as it's called. If we generate an idea, the mind is supposed to send a message to the perception system telling it that the idea is coming in—if that breaks, if the motor schema fails to reach the motor system, either because it is stopped, delayed, or never sent, a hallucination that we aren't the ones moving our own bodies can result. This doesn't involve imagination either, because what's being generated isn't an imagining, but an actual body movement.

Fourth, sensory deprivation can cause hallucination. The sensory system might generate information when it is starved of sensory information from the real world—as people in environments of sensory deprivation can attest to. If the sensory system is starved of information, it will activate itself from expectation, resulting in hallucination. It can happen with sensory loss—blindness, deafness, and so on. This can also happen to people whose senses are just fine, but whose eyes get no light for long time. It can also be more abstract—it might be a kind of social deprivation that causes the feeling of a "third man" on long hikes, or bereavement hallucinations of dead spouses. We expect them around, and when they're not, our minds' expectations go a little haywire. This is sort of like imagination, to the extent that we use imagination to form expectations of what we will see in the world.

Sensory deprivation is thought to be the reason people get "phantom limb," which is the vivid hallucination that your limb is still there (and sometimes in pain), after it has been amputated. Almost all amputees experience this, though over time the phantom limb might "telescope," which means that the phantom gradually turns into just the hand or foot, missing the leg or arm that normally goes in between. Sometimes the phantom limb disappears completely. Brain imaging studies suggest that amputation causes a lack of input from the sensory areas that keep track of your body position (proprioception), resulting in your expectations of what your body feels like taking over. That is, you have a body image that you use to understand and keep track of your body, and when you are not getting sensory input about where, say, your arm is, your body image just makes its best guess as to where the arm must be.[100]

We can look at all dreams as being like hallucinations because they are almost always taken, while we have them, as reality. That is, in dreams, we have no insight (that said, when scientists talk about hallucinations, they normally don't include dreaming). There is evidence that some instances of hallucination happen due to dreams intruding on waking experience, usually when the person is falling asleep or waking up. This fifth type of hallucination is dream intrusion.

The sixth kind of hallucination is caused by low-level sensory areas being activated more or less directly, either with stimulation during an experiment, or from a seizure or migraine. The person might see zigzag lines, because that part of their brain is too active, perhaps because of a seizure. Those areas don't "know" that this activity didn't come from the environment, so it gets interpreted as being in the world. (Of course, insight is often preserved in these cases because the lines violate expectations so much.) These hallucinations don't seem to be caused by imagination.

SYNESTHESIA

The seventh kind of hallucination is synesthesia, which is, roughly speaking, a mixing of the senses. For example, some synesthetics see particular colors when looking at particular numbers. For instance, they might think that the number six is pink, and the number two is black. Not only do they *think* it, but they actually experience the color. When they look at a number six, it appears with a pink halo or tinge. Someone else with number-color synesthesia might have different associations. That is, they don't all agree on what numbers are what colors. I knew an artist who had auditory-color synesthesia. She was a designer, and she could not listen to some kinds of music when working with a painting of a particular color because it wouldn't match. The blue of reggae music would clash with a red painting, for example.

Some have a taste-touch synesthesia ("this chicken is too pointy"). Others have a spatial layout relating to dates, centuries, or days of the week.

Lots of people have synesthesia, and many really like it. Because it doesn't interfere with people's lives, it's more viewed as a "condition" rather than a mental illness. That said, people with mental illnesses of various types are more likely to have it, including psychosis and mood disorders. It can also be triggered by alcohol or other drugs.[101] So we can classify it as a kind of benign hallucination.

To sum up, lots of people hallucinate now and again, and it doesn't necessarily mean that they are mentally ill. Hallucinations happen for many reasons (seven, by my count), and only some of them involve what we'd think of as imagination.

6

Dreaming

Dreaming is perhaps the most vivid example of imagination in human experience in which we enter a multimedia virtual world created by our own minds. Dreaming involves visual, auditory, tactile, motor, and even vestibular imagination.[1] The science of dreaming is a fascinating research field, with far more questions than answers. What happens in the brain when dreaming happens? What is the relationship between dreaming and sleep? Or dreaming and hallucinations? Why do we remember dreams so poorly? Do dreams help us in any way? Why do we dream at all? Scientists have learned some possible answers to these questions, but it remains an active field of research with little that everyone agrees on.

Let's start with sleep. There are two main phases of sleep. One of the phases is called "REM sleep," which stands for rapid eye movement. The other phase goes by the creative name "non-REM."

REM is named after the behavior of the eyes. During REM sleep, the eyes of the sleeper move rapidly. When you watch someone sleep, sometimes you can see their eyes moving and know they are in the REM state. Some birds and almost all mammals (dolphins are an exception) experience REM sleep.[2] The muscles of the body are also quieted, called "muscle atonia." This keeps your body from acting out actions you're doing in your dream—and if that part of the brain is damaged, people do things like sleepwalking.[3]

REM is a light sleep, and the brain state is relatively close to that of waking. If you are awakened during REM, you're more alert than you would be in non-REM sleep. Lots of dreams happen during REM sleep, and one theory of dreaming, called the "scanning hypothesis," holds that the eye movements are actually attempts to attend and focus on fictive images in the dream. Although lots of work has been done on this problem, we still don't know if it's true.[4]

The deeper part of sleep is called "slow-wave sleep," or non-REM (NREM) sleep. If you are awakened during NREM sleep, you're much groggier. Most dreams happen during REM sleep, but there are NREM dreams, too, which tend to be dull, short, more like thoughts, and basically undreamlike.[5] We know this because scientists put people in sleep labs and wake them up at different points in the sleep cycle and ask them to report what they were dreaming about.[6]

REMEMBERING DREAMS

The study of dreaming is plagued with the problem that it is difficult to know when and of what people are dreaming. We forget the majority of our dreams, and, in fact, we also tend to forget waking fantasies—perhaps because they are part of a background mental state.[7] But even with these problems, scientists typically rely on what we can remember about our dreams.

Some people remember more dreams than others. For example, people who have good visual memories and people who are good at visual and spatial reasoning seem to have better dream recall.[8] However, it's difficult to know if this means that they dream more or simply that they are better at remembering them. Although we obviously can't remember the dreams we don't remember, when people report dreaming they often report the strange sense that there was much more to the dream than they can recall.[9]

Children typically are able to report dreams starting from the ages of three to five, and continue to have dreams for the rest of their lives.[10] Again, we have the reporting problem. Do infants or other animals dream? It's difficult to know.

We do know that during sleep, rats form representations of things they've experienced during the day in the part of their brains called the "hippocampus." Is it dreaming? Maybe.[11] We can look at brain activity and find that it is consistent or inconsistent with what we know of dreaming in human adults, and I can watch my pug sleeping and find it very convincing that she's dreaming of running and barking, but without the ability to communicate, we don't know how to get information about their subjective experiences. I asked my pug if she dreamed, but like most pugs, she wasn't particularly forthcoming.

Without conclusive evidence either way, it still seems reasonable that infants might be dreaming. Although we have no access to their conscious experiences when they're sleeping, it's worth noting that we also have no access to their conscious states when they are awake, but we still have no trouble attributing emotions and perceptions to them. Infants spend about 50 percent of their time in REM sleep.[12] They might be dreaming a lot!

Just about everybody dreams, whether they remember dreaming or not. Those who can't have dreams have a condition called "visual anoneria." Interestingly, it is accompanied by a loss of being able to visualize episodes from your past (called "visual irreminiscence").[13]

When people say they don't dream, what's usually going on is that they don't remember the dreams they had. When someone says that they had "so many dreams last night," they might have had just as many as usual, but for whatever reason *remember* more of them. Women seem to remember more dreams than men. We're not sure why, but it could be because women wake up more often during the night (perhaps because they experience more stress), and the more you wake up, the more dreams you tend to remember.[14]

People often remember bits of dreams upon waking, but within a few minutes those memories are gone. One suggested way to learn to remember dreams better is to keep a dream diary. As soon as you wake, write down the dream you had before you forget it. Folk wisdom suggests that this will improve your dream recall because your subconscious mind will pick up the intention of your conscious mind. This might be true, considering that people who keep dream diaries sometimes dream of writing down the dream before they wake—then, when they actually wake, they need to do it again![15] But I can't find any scientific studies to show that keeping a dream diary increases general dream recall.

I had a friend who kept a rigorous dream diary. She was a playwright, and I think she was trying to mine her dreams for artistic inspiration. She got quite good at recall, but found herself wanting to tell people about her dreams. She stopped, though, because it started to interfere with her life. She'd start talking about her dreams, and people would leave the room.

Why are other people's dreams so boring? The flip side of this question is why our own dreams seem so interesting to us. We don't know for sure what the answer is to this question, but I can speculate. If dreams are rehearsals for real life, as I'll discuss later, then talking about them might be adaptive. The idea is that two heads are better than one, and maybe the lessons mined from dreams can be better and more broadly appreciated through discussion.

Another idea is that we like to talk about dreams because most of them are negative, and in general, we have a bias to pay attention to negative emotions. We also pay more attention to unusual things, which accounts for why we prefer to talk about our strange dreams. Another reason might be that we find them so meaningful because they tend to be drenched in emotions—emotions that the people we're talking to didn't have. I'll give you an example.

I once dreamed of a terrifying staircase. I told my girlfriend at the time, and she just laughed at me. In my semi-awake state, I was annoyed with her for not understanding just how scary a staircase could be. She made fun of me for months about that. But our dreams just *feel* important. If you have a negative, emotional, strange dream, it's no wonder you feel the need to talk about it. But because your audience doesn't have the same emotions associated with the dream, they find it so boring.[16]

Anthropologist Donald Symons argued that the reason our dreams are primarily visual and kinesthetic (involving bodily movements) is because our bodies don't need to monitor such things during sleep. It's dark, so you can shut your eyes and don't need to monitor the environment with vision. You should stay still, to keep yourself from getting hurt, so you can safely enjoy motor hallucinations while dreaming. In contrast, you need to be able to hear things, smell fire, and feel pain. If you were dreaming vividly in these modalities, then you might either wake up too often, or not wake when something important was happening in the real world.[17] Although many fail to experience actual pain during dreams, or smell anything, our dreams do seem have a strong auditory component, which is not predicted by this theory. In a survey of memories of dreams, 33 percent of men and 40 percent of women reported having smell or taste in dreams. When one looks at dream diaries, only 1 percent of dreams have these features.[18] This is probably closer to the truth, because dream diaries give us more representative information than memories of dreams, which tend to be biased by what's most memorable.

THE CONTENT OF DREAMS

So what do people dream about? If you want to read descriptions of people's dreams, the website dreambank.net has thousands. Dreambank's database was curated by psychologists Adam Schneider and G. William Domhoff.

One thought might be that we tend to dream about what happens to us. Dreams must be made of memories, so maybe our dreams reflect the experiences we have in waking life. A theme in this book is that imagination, including dreaming, comes from memories. Dreams, then, should be composed of elements experienced before, recombined, sometimes, in creative ways. This is why congenitally blind people don't have visual information in their dreams, only spatial information. People who went blind later in life, however, because they do have visual memories, have the ability to exploit these memories to create visual imagery in dreams for the rest of their lives.

Now there's something interesting called "childhood amnesia": we tend to not have event memories from before the age of about three. So, children who go blind before the age of three should have reduced or no visual information in their dreams, because they cannot access those memories. As far as I know, this data has not been collected. However, children with profound vision loss before the age of four deny having hallucinations, where adults in with vision loss often have them because of Charles Bonnet syndrome.[19]

As imagination is a result of memory retrieval, one possibility is that dreams are just a replaying of plausible versions of waking life. Introduced by Michael Schredl, the continuity hypothesis suggests that dreams reflect waking experience.

It certainly seems that there's a lot to this theory. People dream of the problems they have in real life, such as divorce. Their personalities are intact for their dream-selves. People with stressful lives tend to have more stressful dreams.

What this idea fails to explain is why dreams have significantly more of some kinds of things and significantly less of others, compared to waking life. If dreaming is a mere reflection of waking life, the proportions of waking life should be reflected in our dreams. But they're not.

To take an example of my life, I spend many hours a day in front of a computer screen. I work on this book, papers, I answer emails, play games on my phone, and read books on my Kindle. When I relax, I often watch a movie—on my computer screen. However, I rarely use any kind of computer device in my dreams. Studies show that I'm not alone in this. A study by Schredl found that cognitive activities, especially reading, were much less frequent in dreams than in waking life, where as other activities, such as driving, are overrepresented in dreams. It might seem that we tend to not dream of modern technology, but people do tend to dream of watching TV, using motor vehicles, and talking on the phone.[20]

Dreams also do not necessarily reflect what has happened to us recently, and studies have exposed people to things before going to bed and found that those things are rarely incorporated into dream content. This has been tried with film, visual images, and changes in social milieu. Also, what people do or think about before sleep also has little effect.[21] There is an interesting exception to this: thinking about a particular *problem* before bed can help you dream about it.[22]

Nevertheless, the processes involved in dreaming also seem to be able to access memories that one cannot consciously retrieve. Otto Pötzl ran an experiment in which he showed participants complex pictures, such as landscapes, but displayed them so quickly that they had no conscious awareness of them. He then asked them to try to draw the picture that they had no conscious memory of even seeing. Then they were asked to go home and to have a dream that night. The next day they came in and drew pictures of scenes from their dreams. These drawings contained elements of the subliminally presented picture from the day before that were not in the drawings they made the day before![23] This suggests a few interesting things—that we store, in memory, details that we cannot consciously access and that sometimes these details can be brought to consciousness through dreaming.

But this memory facilitation can happen with imaginative processes unrelated to dreaming, too. Just spending some time fantasizing can help you remember things better than other activities, such as playing darts, according to a study by Matthew Erdelyi.[24]

So there's more to dreaming than just replaying your daily life. So what *do* we dream about?

Dreams are often thought to be bizarre, but, in general, they are not. Most dreams are fairly ordinary, though the dream events are often connected in a discontinuous or illogical way.[25] Careful studies suggest that about 80 percent of our dreams are entirely normal.[26]

This might sound implausible. If dreams are mostly normal, why is it that the dreams we remember are so weird? Why do people believe that dreams are weird, if they're not?

The reason we tend to think of dreams as bizarre is because the bizarre dreams are easier to remember. When thinking about the strangeness of dreams, I often tell of a dream I had in which I was at a funeral of a man who'd insisted that all of his toes be buried in separate, little coffins. It's incredibly creative and interesting—at least to me. We are more likely to tell someone about a bizarre dream, and this rehearsal also increases our memory of it. There is a psychological bias called the "availability heuristic" that makes us more likely to think something is common or probable based on how easily it is brought to mind. Strange and emotionally charged memories are easier to recall. I have told people about this particular dream over and over, so because of the availability heuristic, I am more likely to think of it as representative of dreams, in general. So we end up thinking dreams are weird because when we bring dreams to mind, the weird ones surface.

When dreams are bizarre, we still take them to be really happening during the dream—using the terminology from hallucination, we'd say we have no insight. The reason we don't notice strange things in dreams might be because the dorsolateral prefrontal cortex (DLPFC) and the medial and parietal cortical structures are deactivated. These areas are implicated in what is thought of as executive functions—the decision-making and comparing parts of our brains. The left DLPFC, in particular, is selectively activated during reasoning tasks. With reduced activation of our critical thinking, we are more likely to take normally unbelievable things as real when we dream them.[27]

It is interesting to compare dreaming to psychotic experience. Both involve a break with reality in some way, and they both involve the frontal areas being less active and the emotional (limbic) areas more active. Perhaps someone experiencing a psychotic hallucination is doing something akin to dreaming while awake.[28]

One way that dreams are *not* bizarre is that like our waking experience, dreams are almost always first person, or viewed from inside our own heads, as opposed to like watching ourselves do something on video.

However rare bizarre dreams are, they do indeed happen, so they are still in need of explanation. Dreams must come from memories, so why should *any* dreams be weirder than what we experience in real life?

Many think that dream content is often metaphorical. For example, psychiatrist Erik Goodwyn describes a patient who had repeating dreams of a bear. Sometimes the bear would threaten those she cared about, and sometimes she was concerned about the bear in her dreams. She and Dr. Goodwyn discussed when the bear appeared and when the bear didn't, and how it related to what was going on in her life (getting her first period, getting attention from boys, her first boyfriend, and so on.) Together they concluded that the bear appeared when she felt threatened or taken advantage of. Perhaps it was a guardian? As a result of this discovery, her dreams ceased to be nightmares and she would have positive dreams about the bear. For example, she would dream of resting comfortably with it.[29] It's possible that this interpretation of her dream helped her psychologically. Or, it might be a coincidence and that something else caused the changes she experienced. But let's assume for a moment that this dream interpretation actually was the thing that caused her to improve her well-being. This does not necessarily mean that the interpretation was correct. That is, believing that something in your dream metaphorically represents something does not mean that it actually does, even if that belief helps you. It could be that many other, completely different interpretations would also have helped her.

Unfortunately, it is very difficult to come up with an experimental test of what the bear *actually* represents in her mind. What do dreams really *mean?* We can start by asking what it means for a dream element to have some deeper metaphorical meaning. One way to frame it is that some part of your mind wants to think about something, say, your father, but for whatever reason the father concept is explored by thinking about a related concept like a guardian bear. If there is a causal relationship between the idea of your father and the generation of the bear in the dream, then we can confidently say that the bear, in that instance, represents the father. If this happens in dreams, it would be an unconscious process, but we consciously make metaphors in our waking life all the time. We might describe someone as a lizard or a hen.

Any true meaning of dreams is difficult to figure out, even for the dreamer, because the mind would unconsciously have chosen images to represent other concerns.[30] Was the mind trying to represent something other than a bear when it churned up a bear in her dream? How would one devise an experiment to test it? Currently, the scientific community does not know. It could be that, at some point in the future, we understand imagination well enough that we can show that the processes that generate dreams place metaphorical imagery in place of other ideas. But we are not at that point yet. Although many people feel strongly that their dreams are metaphorical and, indeed, many people enjoy interpreting their dreams in a metaphorical way, we just cannot say for sure, scientifically, whether dreams are metaphorical. But because we can't yet think of a way to find evidence of metaphor in dreams, for the most part, scientists assume that metaphorical thinking does not explain the occasional bizarreness of dreams.

Some data, however, are suggestive. Ernest Hartmann looked at the ten most recently reported dreams before the 9/11 attacks, compared them to reported dreams after the attacks, and found that the dreams afterward tended to have more themes of attack, such as being pursued by monsters or wild animals.[31] If dreams are indeed metaphorical, we have no idea why. Why not just dream about 9/11? Why would the mind code it as something else?

So much for symbolism for a particular person. But what about universal symbols? Some theorists claim that there are very deep symbols that all human beings share, not because of culture, but because we have similar bodies, react to the environment in similar ways, and have the same evolutionary history. For example, Goodwyn provides suggestive evidence that snakes, for example, have particular meanings to people that reappear in many cultures. Snakes seem to be associated with the Earth and being a "material enemy of high-minded heroic gods." He lists over eleven examples of religions in which a snake has this kind of symbolism.[32]

From a scientific point of view, though, theories like this suffer from a lack of what we might call "sampling rigor." For a skeptic to be convinced of a pattern, it is not good enough to merely list examples, because we don't know how many counterexamples are being left out of the list. It's easy to cherry-pick examples that support your theory. If we had a large, representative sample of religions, or symbols in religions, and found that there was a greater than chance frequency of snakes having this kind of meaning, then the theory would be more convincing.

We do seem to have an evolved tendency, shared with other primates, to easily learn fear of snakes and spiders,[33] which makes snakes and spiders prime candidates for cross-cultural, evolved symbols. Goodwyn is on stronger ground when he makes the more modest claim that snakes are generally interpreted as "important" because of our built-in ability to recognize them and easily become frightened of them, resulting in their widespread use in dreams and religions. But more (and better) scholarship is required before we can start attaching meaning to symbols in dreams (or in religion) with any confidence. Unfortunately, the compelling idea of universal symbols will not help us understand either why dreams can be bizarre or what they mean when they are. The books you find that have dictionary-like symbol associations to determine the meaning of your dreams, interesting though they might be, are not based on any science.

Another interesting aspect of dreaming is that it's always animated[34] (as opposed to, say, a series of still images) and that it involves discontinuous scene changes. That is, one might dream of being in one place and then, inexplicably, find oneself in another—something that does not happen in real life. As dreamers, we rarely make any note of this.[35]

Other common dreams include being unable to find a bathroom and one's teeth falling out. We are not sure exactly why we have these dreams, but one theory is that they are based on feedback from our bodies during sleep—a full bladder or tooth grinding, for the teeth falling out dream.[36] Sometimes sensory experiences (even those from inside our own bodies) can get interpreted in the dream. When you need to urinate, but can't because you're sleeping, the dream that would make sense of this feeling would be one in which you can't find a bathroom. I've had many dreams of police sirens that I discovered, upon waking, were caused by my bedside alarm.

DREAMING VERSUS DAYDREAMING

So how similar is dreaming to daydreaming? They differ in a few important ways. In contrast to the dreams we have when asleep, daydreams are much less violent, more reflective, have internal monologue, and are more concerned with planning and life goals. Perhaps the most important difference is that we seem to have much more conscious control over the content of daydreams than we do of night dreams.[37]

In order for us to differentiate between dreaming and *day*dreaming, we first need to understand what happens in the brain when we're dreaming.

During a dream, the brain stem sends input to the newer parts of the brain. One theory of dreaming is the activation-synthesis hypothesis, which holds that this information coming from the brain stem is chaotic—essentially meaningless. Dreaming is the effect of the forebrain (the sensorimotor and limbic regions thereof) attempting to make sense of this chaotic information.[38] That is, dreaming is a cortical interpretation of essentially random events. The fact that the forebrain is involved is why we are "conscious" of dreams. (The word "conscious" gets used in several ways. One meaning highlights the difference between sleeping and waking. This is not the sense used here. In some sense, we are conscious of our dreams even though we are asleep.) The creators of the activation-synthesis idea think that because the dreams are created from chaotic information, the dreams themselves are kind of random and meaningless.

But it is possible that the forebrain takes chaotic information as input from the brain stem and then creates something nonchaotic out of it. The essential part of the theory—that dreams are made *from* chaotic input—does not necessitate that the dreams themselves are meaningless. When I did improvisational theater, we would often take random, unconnected suggestions from the audience and make scenes out of them. But those scenes would be coherent, interesting stories, even though they were inspired by essentially (and, often, intentionally) random input.

So the idea that dreams are interpretations of chaotic input from the brain stem does not require that dreams themselves are chaotic and random—which is good, because dreams are anything but. In many ways, our dreams resemble our experiences in the waking world very closely. Our dream environments are usually very realistic scenes, with the usual laws of physics as we understand them, populated with objects, animals, and other people, woven together into a narrative. Also, the prevalence of having the same dreams over and over, which about 50 percent of people report having, is hard to explain if you consider dreams to be essentially random.[39]

What might actual "chaotic dreaming" look like? Disorders that create noisy activation patterns in the brain, such as migraine auras, hallucinations as you're falling asleep, or Charles Bonnet syndrome hallucinations, do not result in dreamlike narratives, but in more "random" feelings and sensations.[40] We can get clues from these disorders: symptoms of migraine with aura sometimes include seeing zigzag lines, geometric shapes, and colors. Charles Bonnet syndrome includes silent hallucinations of still images of scenery, text, animals, and people.[41] Both of these sets of symptoms are better candidates for experiences made by the mind that are "chaotic," but neither have narrative or other notable aspects of dreaming.

But even zigzag lines and images of walking people are not *completely* random. Compare these to actual visual noise, such as static on a television screen or a radio tuned to nothing. Nobody dreams of static, but if we did we might have evidence that our dreams (as opposed to the input that inspires them) were truly random. As we can see, what we might consider random or chaotic falls on a spectrum from complete noise to fully realistic simulations of the real world. When we look at it this way, dreams tend to fall on the realistic and meaningful end of the spectrum.

Some have suggested a "random activation theory" that suggests that dreams are not random, but are narratives formed by the forebrain to make sense of randomness. The apparent meaningfulness of dreams is a result of the forebrain using memories that are easily accessible—things we care about. In this view, dreaming serves no function at all. It is just a result of processes that are there for other reasons.[42] But even some form of the random activation theory is correct, we are still in need of a theory of *how* the forebrain uses this input to create coherent narrative experiences in dreaming.

Not all scientists agree that brain stem activation is chaotic, however. Neuroscientist Barbara Jones, for example, notes that the brain stem sends out highly ordered patterns of motor movement, such as running, chewing, and other repetitive behaviors. We frequently dream of what are called "programmed movements" in our dreams, such as running. This might be because of the activation from the brain stem, where instruction for these kinds of movements (like those of animals) are stored. The brain stem also generates complex behaviors involving sex and rage, which are also common in dreams.[43]

Dreams are also *emotionally* realistic. That is, if we dream of something frightening, we tend to be frightened in our dream.[44] It's natural to think that we dream of something scary, and this *causes* us to have a frightened emotional reaction. What's interesting is that in an anxiety dream, for example, the dream content might shift from one anxious scene to another. The fact that the emotion stays constant in the face of a changing scenic landscape suggests that the emotion is primary, and the scene is secondary.[45] According to the activation-synthesis hypothesis, the causation is in the other direction. We feel frightened, and the mind constructs appropriate dream plots to fit the emotion!

When dreaming, the more active parts of the brain are those that contribute to visual association (Brodmann areas thirty-seven and nineteen) and paralimbic cortices. The part of your brain thought to be important for the construction of spatial imagery is the inferior parietal lobe, which is also relatively active during dreaming. If this area is damaged, dreaming (or dream recall, anyway) stops.[46] Interestingly, the primary visual cortex, thought to be the seat of visual mental imagery in general, is actually *deactivated* in dreaming. This area is what we use for perception in its early stages. This supports the idea that the imagined sensory information in dreaming happens downstream of the more basic sensory systems. It also might explain why dreams tend to actually have a relatively *low* visual intensity.[47]

As described in the hallucination chapter, we can experience dreamlike experiences under sensory deprivation. One theory holds that, starved of external input, the sensory parts of the mind turn to internal sources for information. This idea views the occurrence of dreams simply as a kind of sensory deprivation during high overall brain activation. A further refinement to this theory is the Activation-Inputs-Mode (AIM) model, which attempts to place every kind of conscious experience as a point in a theoretical space with three axes.[48] The first is brain activation (A). Dreaming happens during high brain activation. The brain is also usually active during waking. The second axis is the origin of inputs (I), which can be either internal or external. Information interpreted as sensory information might be thought of as pseudosensory. This information might come from the brain stem, recalled directly from memory, or formed using bits of memory in newly constructed imaginings. These two axes are relatively easy to understand. The third axis, mode (M), is more complicated: mode refers to the relative level of activation of aminergic (prominent when awake) versus cholinergic (prominent during REM) neuromodulators. These chemicals exert an influence on the brain that helps distinguish REM, NREM, and waking states. Although the brain is active in both REM and waking stages, these states of consciousness differ greatly in terms of mode.

LUCID DREAMING

You might have experienced a state of mind that is not quite dreaming and not quite waking either. Although we tend to believe we are awake during dreams, occasionally, during a dream, we will notice that we are dreaming. The internally generated images we normally take as reality are seen for what they are. This fascinating state is called "lucid dreaming." About 80 percent of people will have a lucid dream at some point in their lives, and it's more frequent in children under sixteen years of age.[49]

Many scientists dismissed the notion that this could even happen, but clever experiments found objective evidence that it's real. Some of the eye movements made during REM correspond to the dreamer's gaze in the dream. Stephen LaBerge took advantage of this and had people make prearranged eye movements during their lucid dreams, like looking left for ten seconds, then looking right for ten seconds. These eye movements were captured with measurement equipment in the real world and distinguished the movements from the normal rapid eye movement you usually see when someone is dreaming. Because the eye movements required conscious effort, this is evidence that the dreamers were aware that they were dreaming.[50] As with other REM dreams, the only parts of the body that can move are the eyes, so scientists use eye movements as a communication channel between dreamer and scientist.

This method also allowed him to figure out if the experienced minutes in lucid dreaming corresponded to actual minutes. It turns out that, at least in lucid dreams, it does (some normal dreams have time compression).[51] With this method, some lucid dreams have been found to last up to fifty minutes, but most are around two minutes long.

If we assume that lucid dreams are, in most ways, similar to actual dreams, then scientists would be able to infer things about dreaming in general from discoveries made in experiments with lucid dreaming. This suggests that experienced time intervals in dreams, in general, also correspond to time on the clock in the waking world. This methodology also allowed LaBerge to determine that breathing in dreams corresponds to the breathing of the body. Although there is muscle atonia during dreaming, movements in dreams cause minor muscle twitches that correspond to dream movement, suggesting that the atonia is not perfect. This is also true for the physiological responses to sexual activity in dreams. I'll spare you the details.

One explanation that has been suggested for what is going on in the brain during lucid dreaming is that the normally deactivated, critically oriented DLPFC is activated more—enough to notice something seems unreal, but not enough to wake us up.[52] This interpretation also makes sense in light of evidence that lucid dreams tend to be more coherent than normal REM dreams. Lucid dreams can also be used for creative endeavors but are less useful for rational problem solving.[53] Daniel Erlacher found that athletes can use targeted lucid dreaming to internalize movements and get better at their sports.[54]

It's interesting that most lucid dreaming is characterized by control of the dream self and, to a lesser extent, the dream environment. That is, in a typical lucid dream, you can control yourself (though I've talked to people who could not), but it's harder to change your dream environment.

Even in lucid dreams the dream characters you meet appear to behave autonomously—as opposed to you having to consciously choose what they do, as you might if you were writing a short story. But lucid dreamers can direct conversations they have with dream characters, which can lead to some pretty entertaining scientific studies. Tadas Stumbrys had lucid dreamers ask dream characters to do math problems. Turns out dream characters are pretty bad at it, performing at a primary school level. For reasons unknown, they are better at multiplication and division than at addition and subtraction![55]

These findings are good examples of how dream characters aren't very good at logical thinking. But they can do a lot of other things. Lucid dream characters can draw pretty well, rhyme words, and even mention a word that the dreamer doesn't know.[56] They are also creative. Stumbrys found that they came up with more creative metaphors than the dreamer could when awake.[57]

I often ask my dream characters questions about things I want to know. I keep a list of questions to ask experts when I meet them. Sometimes, when I'm dreaming (the nonlucid kind), I ask the experts I meet these questions. Unsurprisingly, they are very evasive and don't give me direct answers. I only realize when I wake that I was just talking to myself the whole time, asking myself questions I don't know the answers to.

Lucid dreams are really amazing things, so people have tried to come up with ways to make them happen. Unfortunately, most of the science done trying to show how well these methods work is of poor quality.[58] None of the methods produced reliable effects, though a few look more promising than others, so those are the ones I'll talk about.

The Mnemonic Induction of Lucid Dreams (MILD) technique involves visualizing and rehearsing a dream while you're going to sleep, while focused on the intention to remember that one is dreaming. There is real but weak evidence that this can increase lucid dreaming. It works better in the morning (if you wake up and go back to sleep) than at night. A related technique is the intention technique. Rather than keeping in mind that you're dreaming, you go to sleep with the intention to recognize that you're dreaming once you are.

In the dream reentry technique, you wake from a dream and try to go back into the dream without really falling asleep. You do this by counting or by focusing on your body. Apparently this has its origins in the Tibetan dream yoga tradition.

Reality testing is when you practice testing to see if you're dreaming so often in your waking life that you habitually do it when you're dreaming. When the test fails, you realize you're dreaming. This works because dreams tend to have some reliable characteristics. For one thing, printed words don't often say the same thing when you go back to look at them. So one thing I do is, when I see a sign, I read it, look away, and then read it again. If it says the same thing it did before, I'm not dreaming. The idea is that sometime I'll do this in a dream and go lucid. I have to say that I don't remember many of my dreams, and my lucid dreaming is pretty infrequent. I don't even remember if this technique has ever even worked for me.[59]

I teach about lucid dreaming practice in one of the classes I teach, and one of my students told me that she suffers from depersonalization (feeling that you are not real) and derealization (feeling that the world is not real). She told me that she started doing reality checks to try to start lucid dreaming, but she got the unexpected benefit of it helping with her real-world problems. She just does a reality check whenever she starts to feel that she or the world isn't real, and it helps. She has not had a true episode since.

There are some products that shine light onto your eyes during REM sleep. The idea here is to send a message to your dreaming consciousness. A sudden increase in light is supposed to inform you that you're dreaming.

Again, the studies done in this area are typically poorly done, and none of these methods work very well. There are lots of other methods that seem to work even more poorly, so I'm not going to bother describing them, but if you're hell-bent on lucid dreaming, go ahead and try them.

SO WHY DO WE DREAM AT ALL?

Dreaming is still very mysterious, but now there are several good theories that shed light on what they are and how they work. I've talked about some theories of dreaming that hold that dreams don't mean anything. Many theorists think that dreaming has no function. That is, it's not good for anything. This is possible because there are aspects of human beings (as there are with all living things) that serve no function at all, such as the color of our bones.

However, dreaming certainly seems to be important for something. In a laboratory setting, you can allow people to sleep but prevent them from most of their dreaming by waking them up when they enter REM. This will make them hallucinate in their waking lives, and, over time, they will gain weight, have trouble concentrating, and get anxious and irritable.[60] Alcohol, cannabis, cocaine, nicotine, antidepressants, and even some sleeping pills can interfere with REM sleep, which can also cause a state of dream deprivation. With the use of artificial light at night we've extended the average length of our days by about four hours—this light interferes with our circadian rhythms and dreaming.[61]

We also suspect that dreaming has a function because it's a result of evolution. Dreaming is something people seem to do without learning how, and other animals appear to do it, too. People don't explicitly teach others to dream, we can't observe people dreaming, and it does not appear to happen in response to anything in particular from the environment. These are all characteristics of inborn behaviors. We either evolved *to dream*, which would mean that there was some advantage to doing it, or we evolved to have other properties, of which dreaming is a side effect.

To ask for the function of an unlearned biological trait, such as dreaming, is to ask for an evolutionary explanation. The function of something is thought to be something that helped the species survive and reproduce. Dreaming is an interesting case because it appears to be completely internal.

It seems plausible that we could spend the night with no experiences at all and not be any better or worse off.

Philosopher Owen Flanagan believes that evolution did not select for dreaming, citing as evidence the fact that dreaming doesn't process any actual sensory input and only activates negligible amounts of motor movement.[62] It does not directly get us more food or mates. What could it be doing? But recall that many other mental functions, such as planning and problem solving, also do not rely on our immediate perceptual environment, nor our immediate motor actions. But it would be silly to suggest that our ability to plan does not help us get resources or reproduce.

Sleeping, and REM sleep especially, seems to play an important role in learning. In particular, it is important for procedural learning (for example, the kind of learning you use to drive a stick shift or eat with chopsticks) and visual learning. Other studies suggest that sleep, and even dreaming, is important for visual reasoning tasks as well, such as navigation. However, even REM's role in learning is challenged by some.[63]

Perhaps dreams help you deal with difficult emotional issues in life. It is clear that dreams reflect the current emotional problems of the dreamer.[64] Rosalind Cartwright found that people who were depressed about their impending divorce were more likely to dream about their spouses, and that those who did were better adjusted emotionally than those who didn't.[65] This evidence is suggestive, but doesn't necessarily mean that dreaming *caused* the emotional adjustment. It might be that getting emotionally adjusted resulted in the dreaming, or some third factor caused both.

One clue to what dreaming might be for comes from what kind of content seems to appear in dreams more often than in real life. For example, threatening situations are more common in people's dreams than in their lives. Seventy percent of nightmares are frightening, and the other 30 percent tend to feature other dominant negative emotions, such as frustration, sadness, or anger. Just about everybody has nightmares, though much of the time they, like most dreams, are not remembered. Between 8 and 30 percent of adults surveyed have a nightmare at least once per month—different studies get different numbers. Ignoring the concept of a nightmare altogether, studies of dream emotion find that negative emotions are much more common than positive ones—between

66 and 80 percent. Many bad dreams involve misfortune and aggression. Enemies are common, and if they are human, they are almost always male, regardless of the gender of the dreamer. Dreamers tend to react to enemies by running, escaping, or hiding.[66] The only types of recurring dreams that happens with any frequency are anxious or threatening.[67]

The themes of nightmares are similar worldwide: being chased, attacked, taking tests for which you are not prepared, being paralyzed, being late, being naked in public, and losing people close to you. Strangely, bad dreams tend to happen indoors.

THREAT SIMULATION THEORY

Why do we get nightmares? One explanation is threat simulation theory, championed by cognitive scientist Antti Revonsuo. The idea behind this theory is that some dreams, but not all, are a way for your mind to practice dealing with threatening situations.[68] This practice leads to enhanced performance in the real world, which helped our ancestors deal with the threatening situations they faced. Responses to the common threats of our ancestral environment are represented as behavioral programs, which can get activated by real-world experiences.

An important part of this theory is that, in addition to emotionally charged, recent concerns in people's lives, people tend to dream of *ancestral* threats. That is, things that tended to be threatening during what is called the "environment of evolutionary adaptation"—a long period in the Pleistocene where we were living in bands of about 150 people on the African savanna.[69]

Although we know a bit about how people lived back then, thanks to findings in archaeology, most of what we believe about their lives comes from studies of contemporary hunter-gatherer societies. We assume that the lives of our ancestors looked a lot like those of contemporary peoples living in similar environments, with similarly low levels of technology. The evolutionary explanation is that those who could *virtually* practice dealing with threats, in their dreams, would have an advantage over those who learned to deal with threats *only* from real life—which would often be a deadly learning experience: the major causes of death for our ancestors were likely predation by animals, infectious disease, exposure to the elements, risky activities during hunting and gathering, and violence with other people.[70] Although it might be true that some people survive just fine without being able to dream, the important question is whether, over time, dreamers would have any kind of reproductive advantage. By analogy, just because some people can't run (due to knee problems, for instance) and still manage to survive and reproduce, doesn't mean that running, in general, is not beneficial to survival.[71]

When we consider this, some of the common themes of frightening dreams start to make sense. Recall that threatening encounters with men are much more likely in the dreaming world than in the waking one,[72] and that most human dream enemies are male. In our contemporary society, everyday encounters with strange men are not typically threatening. However, in our ancestral environment violence was probably much more common, and intergroup violence was (and still is) almost entirely conducted by males.[73]

We also tend to have a proportion of dreams about dangerous animals that far exceeds what we would expect given our daily experiences. Think about how often people have nightmares about spiders and snakes compared to how often, in their real lives, they are actually threatened by them. This suggests that the fear of snakes and spiders (or at least the propensity for that fear) is a part of our genetic makeup (studies with monkeys support this).[74] This also explains why phobias also tend to be about ancestral dangers—you're more likely to get killed by a car than a snake, but car phobias are virtually unheard of. How often, in the Western industrialized world, do we find ourselves being chased by malevolent creatures or people? Yet being chased is one of the most common dream scenarios. You might think that this is because we see a lot of movies in which people get chased, but this pattern was also true in the 1940s, when people were exposed to much less media than people are today.[75]

Further support for the genetic component of animal fears comes from the observation that children tend to dream of aggression and animal danger the more than adults, and that this tapers of as they mature. If fear of animals were learned in the environment, we'd expect the opposite trend.

Further, children's dreams tend to be more aggressive and are more likely to contain animals and conflict. The animals they dream of—like gorillas and wolves—are almost never encountered in the waking world,[76] except for maybe at zoos. And it has to be an unusually bad day at the zoo if the child somehow gets chased by one.

If the child grows up in a world relatively free of dangers, then these kinds of dreams decrease over time. This suggests that frightening dreams are evolved.[77] For threat simulation theory, the reason we don't tend to dream about reading and writing is because these activities have only recently become part of our culture (thousands of years is recent on evolutionary time scales).

Children's stories and fairy tales contain lots of animals—couldn't that be the reason why they dream of animals so much? Possibly, but there are a couple things we should keep in mind. The first is that fairy tales often involve anthropomorphized animals that talk and interact with human characters like other people do. However, children's dreams tend to feature fairly realistic animals. This casts doubt on the idea that stories are causing the high animal content of dreams.[78] The second thing to keep in mind is that the fact that kids are so interested in animals is a mysterious phenomenon itself. Why are children so fascinated with animals and dinosaurs, in a way that most people seem to, for lack of a better word, outgrow? It could be that children dream about and want stories about animals for the same underlying reason: that they are preprogrammed by evolution to pay close attention to them.

The idea that children's stories affect their dreams is a part of a larger idea that dreams are influenced by modern media. An analysis of dream content found that 63 percent of threatening dream content (of university students) was content that could, in principle, be encountered in the personal life of the dreamer. The media explained about 33 percent of the threatening dreams, and explicit fiction and fantasy accounted for only 4 percent.[79]

Dreams would not be particularly useful as a kind of virtual practice if we did not behave appropriately in them. But we do. An analysis of dream content showed that 94 percent of reactions to dream events are appropriate. The remaining 6 percent were physically impossible or irrelevant to the threat dreamed of. Sometimes, when people dream of threats, they find themselves unable to move in the dream. This, too, is relevant, as the "freezing response" has been found to be useful to avoid being noticed by a predator in many animals.[80]

Earlier I mentioned how difficult it is for laboratory experiments to get presleep stimuli to show up in dreams. From these findings, it seems reasonable to conclude that what happens in your life has little effect on your dreams. But there is one very glaring exception: people who experience trauma in their lives tend to have more nightmares about those experiences.[81] This is further support for the threat simulation theory.

One speculative theory holds that traumatic experiences actually stimulate the development of the dreaming mechanism. Children who have experienced trauma report more fully developed nightmares than other children. Maybe children who have no trauma are slower to develop dreams at all.[82]

PTSD can cause nightmares that plague the dreamer for a very long time—sometimes as long as fifty years. It is hard to believe that this is doing the dreamer any good, as threat simulation theory might suggest. These nightmares are debilitating.[83] Keep in mind that threat simulation theory holds that dreaming was helpful for our ancestors by making them better at survival—this does not mean that they were made happier, only that when facing situations resembling those they dreamed about, they would be better prepared for them. Further, modern hunter-gatherers, though they frequently experience threats, do not often get PTSD, which requires prolonged experiences of terror that are rare in a hunter-gatherer society.[84]

Threat simulation theory sees dreaming as a kind of practice, to prepare yourself for real-world events. Take, for example, a dream of being chased. The dreamer has a vivid experience of running. If this simulated practice is to be of any use, then dreaming of moving your body should use the same brain areas as actually moving your body. This is exactly what happens. Motor imagery implicates the same brain areas as actually moving.[85] Muscle atonia prevents the signals from actually stimulating the muscles in our body. The communication lines are cut, keeping us from acting out in our dreams. There are disorders in which this cutting of the communication is compromised, and people act out their dream motor imagery and can be a physical danger to themselves and others. There is a disorder called "REM Sleep Behavior Disorder" in which the normal quieting of the muscles is compromised, and people act out what should be only motor imagery in their dreams, sometimes hurting themselves.[86]

This is why virtual practice, in dreams and in the imagination, might make one more effective at acting in the real world. It rehearses how to respond, both in high-level decision-making, as well as individual muscle movements.

Threat simulation is not the only common thing in dreams. Although many of us might remember dreams of flying, they only account for about 1 percent of dreams. Sexual activity accounts for 4 percent of female and 12 percent of male dreams, but often even sexual dreams are fraught with negative emotion.[87] Social interaction in dreams is much more common than it is in real life, suggesting a theory that dreams evolved to help social bonding.[88] These social dreams are more challenging than social situations in waking life; Revonsuo has moved on from threat simulation to his "social simulation theory," suggesting that dreams are also (or instead) for practicing social skills.[89]

All of this suggests that dreaming might serve a function of general practice. It just so happens that many of the kinds of practice we need are for threatening situations. This broader social simulation theory covers more dream content but does not make as specific predictions about dream content as the threat simulation theory. Dreams, and imagination in general, allow for the experience of exaggerated activity that might be dangerous in real life. You can practice extreme maneuvers and extreme emotions in a safe, virtual environment.[90] It might be scary, but you're not going to get hurt. And evolution doesn't care if you're terrified, as long as you have kids.

In support of all of these practice theories, we see that the dream self is usually quite realistic, as it should be for effective practice. It's the other dream characters that are more often different. Usually they look the same—90 percent of the time, the differences are in behaviors. That is, the dream characters act differently than they do in real life. This might be the mind's way of exploring possible future events that you might have to deal with.

DREAM THEORY AND MIRRORING REALITY

Recall that when we are actively perceiving the world, we are constantly making predictions about it. This makes us faster at perceiving things. Our minds are constantly churning up expectations about what we're about to experience, and what we actually sense then constrains and corrects it. Allan Hobson and his colleagues suggest that dreaming is this expectation system running without sensory supervision. That is, when we dream, it's just like how the sensory prediction system works when we're awake—the only difference is that the senses are not there to correct it—and recall that this is also a theory of hallucinations caused by sensory deprivation. In this view, the function of dreaming is to simplify and understand our sensory world by reducing the complexity of our experience—that is, finding patterns. It's an interesting compliment to the simulation theory I talked about earlier—rather than focusing on what to do in hypothetical situations, this theory focuses on honing what to expect in our sensory world when awake.[91]

This theory would predict that people who don't dream would suffer some kinds of cognitive problems. There are people who can't dream, but there have not been any studies to see if they suffer the cognitive deficits predicted by theories like Hobson's.[92]

When we dream of running, it doesn't just look like we're running, it *feels* like we're running. When awake, we use proprioception to know the positions of our bodies and how they move. But when we're dreaming, our muscles aren't moving—so how is it that during dreaming we feel as though we are? It appears that we know how our bodies are moving not only through the perception of how our muscles and limbs are orientated, but also by the commands we send to them.

In the hallucination chapter, I told you about the experiment with paralyzed eyes, showing that your perceptual system unquestioningly uses the commands from the motor system to know where the eyes are looking, ignoring where they actually are looking. Something akin to this might be happening in dreaming, too. The motor imagination must be sending information to the proprioceptive system regarding the imagined movements to fool it into thinking that the body is actually moving. This way, the illusion that the dream is real can be maintained, and there is no interference from where your body parts actually are in the real world.[93]

This view of dreams as practice also suggests a reason why we take dreams to be true when we have them: if we knew they were dreams, we might be less motivated to get ourselves out of trouble in the dream world.[94] By dreaming of danger, and believing we are in it, our minds get a safe simulation in which to practice with all of the motivation for success that we would experience in real life.

On the one hand, this seems reasonable, if you accept threat simulation theory. If something bad happens to you, you might want to prepare for it happening again. However, for many people whose nightmares of traumatic past events recur, the same terror is reexperienced night after night, so that they fear sleep.

But people who watch scary movies do, too—what is probably happening is that the more primitive parts of our minds interpret the movie as reality, and set up dream scenarios to practice what we will do when we encounter the situations we've seen others in. Michael Schredl found that nightmares tended to reflect the tropes of popular culture. He found that nightmares of the 1920s involved the bogeyman, while nightmares of the 1950s and 1960s featured ghosts, witches, and devils. In the 1990s, nightmares were full of movie villains.

If dreams, or nightmares at least, are supposed to be practice for what are perceived to be real-world dangers, then some aspects of nightmares aren't obviously consistent with the theory. For example, one common nightmare involves your teeth falling out. This is a scenario almost never seen in real life nor in movies. Also, it's something we can't do anything about should it happen (one study showed that tooth-related dreams correlated with dental irritation).[95] Further, some dreams are just too weird to be relevant to our lives, like my dream of a terrifying staircase. Some believe that even stranger dreams are metaphorical representations of anxieties that are difficult to grapple with. Barry Krakow studied the dreams of 168 women who had experienced sexual assault—many of their subsequent nightmares were not clearly replays of the event.

When athletes get better at their sports through mental practice, they are usually visualizing an idealized version of what they're trying to do. But in nightmares, the threat is eliminated in only 17 percent of dreams. If we were practicing to overcome these difficulties, you'd think the percentage would be higher.

Another theory of nightmares was proposed by sleep researcher Ross Levin: fear memory extinction. According to Levin, nightmares expose people to things they fear in a safer, less frightening context. Just as some phobia treatments involve exposing people to what they fear in a safe environment, perhaps our nightmares allow us to get used to things that scare us, because nightmares can have elements that are not frightening. For Levin, the system can break down when the dreamer is under extreme stress, and this can result in repeated nightmares that don't eliminate fear over time.

Nightmares are examples of involuntary imagination, and to some extent, they are affected by our experiences in the real world. As usual, deliberate imagination can be used to alter the mind's interpretation of the world by providing rich virtual experiences. Krakow asked the women in his study to write down their nightmares and to change them to make them more pleasant. He had them spend between five and twenty minutes every day imagining the revised dream. After six months they had fewer disturbing dreams and were sleeping better (relative to a group who was not given the treatment). Similar treatments are being used for other victims of trauma, such as war veterans with PTSD.

The traditional view of nightmares is that they are symptoms of some mental distress, and that the best treatment for them is to treat the mental distress itself, be it through therapy or psychoactive drugs, such as prazosin, which seems to attenuate nightmares as long as the patient is taking it. The idea that nightmares are there to help us, either through practice or because of habituation, would suggest that simply removing nightmares from people's lives might do them a disservice. However, this does not appear to be the case. People who have problems with nightmares get better sleep and are healthier when they can get their nightmares to stop.[96]

Two of the main theories of dreaming we've discussed—that dreams are created from chaotic input from the brain stem and that dreams serve as virtual practice—are actually compatible. I did theatrical improvisation for over twenty years, and we can think of what the frontal areas are doing as an improv show, where the brain stem acts as the audience, throwing out suggestions. According to the activation-synthesis hypothesis, this is what the frontal areas are doing: constructing narratives to make some kind of sense of information from the brain stem. Let's see a scene using the suggestions of the color purple, an elephant, and Natalie Portman. Cognitive scientist Rita Ardito suggests that without dreaming, the chaotic input from the brain stem would prevent us from sleeping by keeping our brains in a psychotic state. On this suggestion, the adaptive function of dreaming is to make stories out of the random stimulation so that the brain can do what it needs to do when it's sleeping, such as restoring itself and consolidating memories.[97]

What threat simulation and social simulation theories are saying is that there is a dark theme to how these suggestions are interpreted. You can make any kind of scene you want from any suggestions, and according to threat simulation theory, the mind takes these opportunities to practice dangerous situations. Rather than a happy scene with an elephant, your mind makes a scene in which it's chasing you.

Your dreams are like performances put on by an anxious improv team. It is your unconscious mind doing its own form of imagining.

7

Mind-Wandering and Daydreaming

Now that we understand what's happening in our brains when we dream, we can turn to daydreaming. Daydreaming is a mental departure from your current experience of the outside world. Often a daydream is fanciful in some way, and unrelated to whatever activity you are currently engaged in in the real world. For example, you might be sitting in a lecture (not one of my lectures, but some boring one) and find your mind wandering. Jonathan Smallwood differentiates two different kinds of mind wandering. "Zoning out" is unintentional. You are kind of washed away by the daydream. "Tuning out," in contrast, is deliberate, and you keep one foot in reality.[1]

How long do daydreams last? Eric Klinger ran a study in which he trained a group of participants to estimate the duration of their thoughts, daydreams or otherwise. What he found was that the average thought length was fourteen seconds, but there were more thoughts that were about five seconds long than there were of any other length. That the mean is so much higher than the median suggests that we have some very long daydreams that bring the average up so high. Every single day, a typical person entertains about four thousand thoughts during their waking hours. If we use, as a working definition of daydreaming, that it consists of thoughts that are either spontaneous or fanciful, then fully half of these thoughts are daydreams—two thousand per day![2]

Daydreams are typically spontaneous, in that you often don't choose to daydream, you might realize you are doing it sometime after it starts.[3] Spontaneous daydreaming is sometimes called "mind-wandering." This often happens when you are bored or doing well-practiced tasks, such as driving, which don't require a great deal of conscious attention. Because your "task at hand" is easy, the mind has processing resources in excess of what's needed. The mind is basically at rest at these times, and to occupy itself it turns to thoughts that are unrelated to the task at hand. People daydream more when stressed, bored, sleepy, or in chaotic environments. They daydream less when enjoying themselves.[4] They might plan for the future, fantasize, or worry about things unrelated to where they are and what they are doing.

In these cases, the metabolic demands of the task at hand turn out to be only a fraction of the metabolic energy consumed by the brain. That is, the thing it looks like someone's doing (say, walking while carrying a box of stuff) is only using a little bit of the brain's resources. This means that a good deal of energy is being used for something other than actively engaging with the external environment. And this is a lot of energy—a full 20 percent of the body's energy is used by the brain, even when the body is at rest. The brain is a mere 2 percent of the body, by weight.

Why is the brain so hungry? What is the brain doing with all of this unused processing power? One theory is that it's trying to make sense of what's already happened. Arielle Tambini had people try to learn pictures of faces and scenes. After the task was over, the brain areas that were important for remembering faces and scenes were still pretty active—more active than other areas that would have been important for different tasks.[5]

At least some of mind-wandering is done with the mind's default mode network. Several brain areas are implicated in the default network. The medial prefrontal cortex helps us imagine ourselves and the mental states of others. The posterior cingulate cortex is for retrieving personal memories, and the parietal cortex is closely connected to the hippocampus, which is used for a lot of retrieval of memories of events.[6] A brain imaging study by

Malia Mason found that people with the most activation in this default network also had the highest incidences of daydreaming.[7]

WHAT'S IN A DAYDREAM?

So what do people tend to think about when they are mind-wandering? A study by Benjamin Baird found that people think about their future more than anything else.[8] Prototypical daydreaming involves visual imagery, but it need not. Much daydreaming involves self-talk, which is experienced as talking to yourself in your head, rather than out loud (74 percent of daydreams involve sound or words).[9]

Daydreaming gets a bad rap, and there's a lot of evidence to support its detrimental effects. Although they use similar parts of the brain, anxiously worrying, or reviewing mistakes made in the past, is different from the mind-wandering I've been talking about. Worry, regret, and anxiety can be difficult to switch off. Positive distractions, such as exercise and being social, can help these ruminators distract themselves from their unpleasant imaginings.[10]

As you might imagine, pressing, highly emotional concerns, such as the fear of losing your lover, can actually interfere with whatever task is at hand. Here is where the imagination can get in the way of your day-to-day life. If you're worried about a breakup, or that Disney might stop making Star Wars movies, you might not be paying as much attention as you should when you cross a busy street, for example. Many studies show that mind-wandering negatively impacts performance on reading, maintaining attention, your ability to prevent automatic responses, as well as general working memory and intelligence.[11]

DOODLING

Shockingly, sometimes being *slightly* distracted from your current task can sometimes help. Jackie Andrade had two groups of students listen to a boring lecture and tested them on the lecture's contents afterward. One group had to just sit there and listen, and the other was allowed to doodle. In spite of all of the evidence we have that multitasking doesn't work, and that secondary tasks make us perform more poorly on primary tasks, she found that the doodlers had *better* memory retention than the other group.[12]

Why on Earth?

Let's think about what it means to be bored. Humans are curious creatures who like to be occupied with moderately challenging situations. Our minds are hungry for information. When what we are experiencing does not engage us, our minds have unused resources. Think of it as mental energy that is looking to be spent, analogous to when your body is antsy, but you're required to sit still. What does the mind do in this situation? Well, it often engages in mind-wandering. The students who were not allowed to doodle got bored, and then completely disengaged with the lecture, thinking about different things. As a result, they retained very little. Their minds were elsewhere.

But the doodlers had an advantage. One way to think about it is that the lecture didn't satisfy their minds, but the combination of doodling and listening to the lecture did. The sum of the two activities kept them from disengaging completely, and they were able to retain more of the lecture![13] (I never criticize my students for doodling during my lectures.)

There's a similar finding for how pilots mind-wander in cockpits.

Cockpits used to be busy places. With modern technology, however, the autopilot does most of the work. Why would we make planes like this? The idea is that with less to do, the less the pilots can screw something up. They will have more attentional resources to devote to any problems that arise. But with all of that automation, do pilots' minds wander off task for lack of anything to do? Do they, like the students who don't doodle, completely disengage from the job of flying the plane?

Stephen Casner put pilots in a flight simulator and used two different levels of automation: one low and one high. Would increasing the amount of automation also increase how often the pilot was thinking about something else entirely?

He asked the pilots at various times to categorize their current thoughts as either specific to the task at hand, higher-level thinking that was still related to flying (such as planning ahead), or thoughts completely unrelated. Luckily for anybody who flies, he found that 83 percent of the time the pilots were thinking about flying, at either level of automation. And as we'd expect, the group in the more automated simulator had more high-level thoughts about flying than low-level thoughts about the tasks at hand—the moment-by-moment issues of the plane. But they also had fewer flight-related thoughts at all. Especially when everything was going according to plan, the pilots often thought about something else. So it seems that automation in the cockpit might be a mixed blessing. It simultaneously allows the pilots to think more strategically about the flight, without having to sweat the details, but at the same time makes their minds wander more.[14]

This kind of thing should be familiar to anybody who has learned how to drive. Think back to when you were first learning—there seemed to be so many things to do, it felt impossible to keep track of all of them: release the clutch, now break, steer to give the cyclist plenty of room, rewind the tape so you can listen to *Rhymin & Stealin* again. There's a lot to do. With practice, though, it got manageable. This is because more and more of driving became an unconscious process, automated, to allow your mind to concern itself with higher-level aspects of the driving experience. After driving for twenty years, you can drive home and think about other things entirely, and often not even have a memory of the drive home at all! Your mind was elsewhere.

When your conscious mind is really engaged, mind-wandering doesn't happen.

THE PURPOSE OF DAYDREAMS

So what are these daydreams for? One clue is that they are less bizarre and more coherent than the kinds of dreams we have at night.[15] One important function of daydreaming seems to be to help us refocus on our long-term goals. When we initially form goals, we commit to them, but we often cannot instantly achieve them. For example, we might have a goal to plant a garden. To successfully get a garden going, one needs to do many steps at different times. There are dependencies—you can't plant seeds before you buy them, for instance. How do we keep from forgetting about the goal? It appears that one function of daydreaming is to reinforce and remind us of these long-term goals.

Suppose one night, in bed, Emily decides that she wants to grow a vegetable garden. The next day it does not come to mind until she is talking to a friend, who mentions that he is growing lettuce. This reminds Emily of her own goal, and as she is walking with her friend she starts daydreaming about her garden—what it will eventually look like, that she has to go buy seeds, that she really should weed the patch of dirt in her back yard, and so on. This daydreaming can be interrupted by something in her perceptual experience—the approach of a dog she really ought to pet, for example—or she might explore the idea in her daydream sufficiently such that she no longer feels the need to think about it anymore. Without daydreaming, her goal might have been completely forgotten. Daydreaming reinforces our commitment to achieving our goals and serves as a reminder of our larger agenda. It can act as a virtual rehearsal, and an opportunity to learn from past mistakes by virtually experiencing them.[16]

There is also evidence to suggest that finding yourself unable to achieve your goals will result in reduced daydreaming about them, as though your mind doesn't want to waste time thinking about things you can't have. For example, a study by Barbara Bokhour found that prostate surgery patients, who have impaired sexual function, then experienced a reduction in sexual daydreams.[17]

If daydreaming and imagination are thought to be used to keep goals in mind, does it work? Does thinking about your goals help you actually achieve them? The answer to this question depends on *how* you think about your goals. Recall that the work by psychologist Shelley Taylor, who found that if your imagination focuses not only on the happiness you'll feel when you've reached your goal, but also on the challenges and steps required to reach it, your chance of actually attaining the goal will be increased. On the other hand, if you focus only on the pleasure associated with goal achievement, it can hinder your ability to actually reach it! Planning on how to get a goal is more effective than merely thinking about the outcome.[18]

It seems like the expansive nature of mind-wandering might help people be more creative. There is some evidence for this. Psychologist Benjamin

Baird was interested in how mind-wandering could help people be creative. He measured creativity using the "unusual uses task," which asks people to come up with as many uses for an everyday object as they can. For example, a brick can be used as a paperweight or to sharpen a pencil. The more uses people come up with, the higher their measured creativity score. Baird had people do a bit of the unusual uses task, and then split them up. Different groups of people engaged in different activities before continuing the test: engaging in a demanding task, resting, not having a break at all, or engaging in an undemanding task. The group that worked on an undemanding task, which encouraged mind-wandering, had the most improvement on the later portion of the unusual uses task. This is because undemanding tasks, such as folding laundry, allows your mind to wander. In fact, he measured the amount of mind-wandering that happened and found that the more there was, the better the improvement. Interestingly, he also measured how many thoughts were explicitly directed to the unusual uses task, and this wasn't predictive of good performance on the unusual uses task at all! This is an example of "incubation," where a bit of quiet time, and mind-wandering, can increase creativity, even for things you're not consciously thinking about.[19]

INTERNAL VERSUS EXTERNAL STIMULATION

So though imagination can be triggered by external forces, such as when you're reading a novel and picture what's happening, when your mind is wandering, it's typically some internal trigger that causes the imagination. We seem to have equal amounts internally triggered and externally triggered thoughts (about 6–7 percent of our time in each case).

Not all imaginings make you happy. During mind-wandering, about half of what we imagine is unpleasant. Unintentional reverie appears to be more enjoyable, though we don't know why.[20]

When we think of using our imagination, we often think of fantasizing. Sometimes it's fun to fantasize about things, but in general people don't like to be left to their own thoughts. Timothy Wilson ran eleven studies in which he asked people to sit in a room by themselves with nothing to do but think, for about six to fifteen minutes. People hated it. In fact, they preferred to do mundane tasks, and even give themselves mild electric shocks, rather than be alone with their thoughts.[21]

And solitary confinement is so terrible that it's the punishment they give in prisons—a punishment for people *who are already in the midst of another punishment*. Prolonged social isolation can lead to a bunch of bad psychological outcomes, including cognitive and spatial distortions and suicidal tendencies. It's so bad that some argue that solitary confinement is a violation of human rights.[22]

As bad as isolation is, some people learn to deal with it through extensive use of their imaginations. A man named Hussain Al-Shahristani spent a decade in solitary confinement at Abu Ghraib prison. He entertained himself by making up mathematical problems for himself to solve. Jakow Trachtenberg invented a new way to do math while in a Nazi concentration camp. Others make up fantasy stories for themselves, or try to replay movies in their heads, or create alternate realities.[23] This can be hard to do, particularly when battling the depression that often accompanies isolation, but it makes the time more tolerable. These Herculean efforts aside, most people hate being left alone with their own thoughts, and if they have to do it for too long, they might develop a mental illness.

But that's most of us.

For some people, their mental fantasy life is so profoundly wonderful that they imagine *too much*. The person feels constantly tempted to enter their imaginary world rather than participating in the real one.[24] Though they find daydreaming fulfilling and pleasant in the moment, their inability to stop engaging in it makes it comparable to an addiction. It's usually called "maladaptive daydreaming," but I prefer the term "compulsive fantasizing," because daydreaming is a broader category, and this phenomenon is specifically fantasizing: people with this problem are particularly engaged in creating fantasies, as opposed to other kinds of daydreaming, such as thinking about how to solve a problem, going over what one needs to do, or reliving past experiences.

Jayne Bigelsen sent out questionnaires to people having problems with excessive fantasizing.[25] She found that, in general, the fantasies were very vivid, often spanning generations and continents, with deaths, marriages, murders, and reunited lovers. Often characters will be borrowed from media, such as movies, but will also include people from real life. Sometimes the fantasizers will do extensive research on historical periods to help inform the fantasies. Although the fantasies can be in any genre, if you will, from science fiction to romance, the most common fantasies are generally character-driven and often aspirational, self-oriented daydreams, such as fantasizing about being a brain surgeon.

It's not an official disorder in the psychiatry manual (the *Diagnostic and Statistical Manual of Mental Disorders*, fifth edition, or DSM-V), at least not yet, but lots of testimony from people suggests it's a real problem for them. One person who suffers from it said, "Sometimes I wonder if it's destroying my life completely. Sometimes I think it's the best part of my life."[26] These people are the best examples of people with too much imagination—so much that they are lost in the creations of their own minds.

Maladaptive daydreamers have trouble focusing on books or conversations. If they get the slightest bit bored with anything, they can sneak away to their own fantasy worlds, which are reliably entertaining. Some people do it for hours a night, some binge for days at a time. Some have described it as being like an alcoholic with an unlimited supply of booze at hand all the time.[27]

After I talked about compulsive fantasizing in my class, a student came to me and said that he had this problem. He said it was like there was a really good TV show on in the room all the time, and you're constantly tempted to watch it. Sometimes these daydreams are replaying negative things that have happened, and they are unpleasant. When this happens he can sometimes "change the channel" to a different fantasy, but he can never shut the daydream off completely. It's always in the back of his mind, distracting and prominent, trying to get his attention.

He could even fantasize while he was doing something intense. "Downhill skiing was one of those things where I had so much fun with it that my brain was kind of like, all right, I gotta focus on this because this is too much fun. Whereas basketball there were sometimes I wanted to play and sometimes I didn't, and when I didn't want to play, I'd play while daydreaming."

I asked him about how he'd fare in solitary confinement. He told me: "If you did it sometime between the ages of . . . I don't know, if you sat me in solitary confinement between six and twenty, I might have been able to."

At that time, he would have been happy, just sitting there in solitary confinement, fantasizing all day long. For now, "fourteen hours might be pushing it. I could do a solid eight."

He has about as much control over the fantasy story as he has over a lucid dream. The characters are autonomous. "I set up a main storyline, and then once the story starts going, it just kind of continues on from there by itself. If I want to control it. If I don't, then it's one of those things where it's . . . it's random."

He could will a fantasy into place—like a cyberpunk story. He sets up the main storyline, and then his subconscious can just run the story, influenced by what he last focused on, such as a television show he recently saw. The same story keeps replaying, over and over, but changes and resets itself.

Interestingly, he doesn't seem to have particularly vivid visual imagery! If he closes his eyes and just lies there, the fantasies are pretty vivid, but when normally awake, not so much. "It's less shapes and more actions."

The fantasies sometimes keep him up at night: "I usually give myself a half hour to forty-five minutes to get to sleep just because of that."

But he still wants to learn to lucid dream, because a fantasy, as fun as it is, "It still feels like it's just an illusion. Whereas when I'm dreaming, everything is more real."

He told me that he'd learned tricks to help manage the problem. "I learned how to multitask. . . . If I feel like daydreaming, then I'll spend more time on the daydream. And then, if I feel like doing something else, I'll kind of just slowly push the daydream back until I'm focusing more. But they're both still happening at the same time."

But to learn to stop it completely, he had to do dangerous, all-consuming activities. "The second trick I learned was, when I'm learning something new, or doing something dangerous, that's when it tends to shut off. So three things that I did to help get rid of it were learning how to skydive, learning how to ride a motorcycle, and taking a job on the rigs."

So he deliberately took a dangerous job on oil rigs just so he could feel what it was like for the fantasies to stop. Apparently, understanding the feeling of what it was like to have the fantasy not there helped him be able to stop them at will, in nondangerous situations: "It's one of those things where once you learn how to do something, you can do it in the future. So when you know the feeling of okay, this is pushing it back, this is, this what it feels like to have to focus, I kind of got a grasp for what it felt like, and needed to do, and it helped me with the future."

But in general he can't shut it off; he can just give it a little less attention, like trying to ignore a TV in the room that's always on. He's had to learn ways to deal with this, but the good part is that he doesn't suffer when he's not able to engage in his fantasies, like many compulsive fantasizers do: "Since it's always playing, I don't get that anxiety or grumpiness. Not withdrawal, because it always seems to be there."

I asked him if he'd could get rid of it, would he? "I have talked to therapists, and they all—they never really diagnose it or anything. They're just like 'Oh you just have an imagination; everyone has it, just do your best to stay in the moment' kind of thing . . . I wouldn't get rid of it completely. There are times when it's nice to just sit back, relax, and enjoy a show, even if it's not on TV. Um, but there are times when I wish, okay, where it's like, well if I could get rid of it completely for a short period of time, like what could I do? . . . Like what more could I do? Could I learn faster?"

One problem maladaptive daydreamers have is that the people they interact with in their imaginations can be more interesting than the people in their actual lives.[28] Nine percent claimed to have no friends or close relationships with real people—but claimed to have meaningful conversations with the people in their heads. I spoke on the phone to one woman who said it was hard for her to stay satisfied with her husband because her dream lovers were so much better.

Of people who suffer from it, 57 percent reported that the fantasizing is done so much that real-world responsibilities are neglected. Some have trouble sleeping because their own fantasies keep them awake. Respondents reported an average of 56 percent of waking hours were spent fantasizing, but in individuals the amount of time varies greatly—generally from one to ten hours per day.

Often this fantasizing is accompanied by some regular movement, usually pacing. Those who *need* to do the movement often will turn down social engagements to stay home and fantasize. This is kind of a benefit, because if they are in a social situation where they can't do the movement, like out at a restaurant with friends, the fantasies can't intrude. But for people who don't need to move in any particular way, fantasies are a constant temptation. The fantasy is always there, and their attention is often divided, leading to exhaustion and an inability to concentrate on anything other than the fantasy.

When real-world responsibilities become pressing, most can suppress the fantasizing, but there is a cost. Anxiety and stress builds as time passes. One person reported that after studying for final exams, they needed to fantasize for three or four days straight. The person feigned sickness to avoid other people and questions. Another actually feels illness—head and stomach aches—if they don't indulge in the fantasies. In one case study, a patient viewed real-world obligations as things that needed to be accommodated so that she could reenter her imaginary life. Think about that: living life in the real world was like the rent she paid to be able to live in the fantasy life she was really interested in. Like everyone else, the daydreaming was less likely during engaging external activities, such as acting in plays. But she easily slipped into fantasy when alone, or on walks, or when conversations with friends got boring.[29] (If you think you might suffer from compulsive fantasizing, there are online forums, such as wildminds.ning.com, where you can get support.)

One might think that for a creative person, such as a novelist or screenwriter, being able to effortlessly come up with plots and characters would be a great benefit. But for people who fantasize excessively, the ideas come too fast to write down and develop. After all, the idea is only the first part of writing a novel or screenplay or any form of creative endeavor. Although 46 percent write stories and poetry, they still want to control their fantasizing so they can have the mental space, time, and attention needed to make something worthwhile in reality. Writing something good, something that others will appreciate, requires care and rewriting. For these people, the story in the head never stops, and their writing can't keep up, let alone make edits.

MINDFULNESS

Most of us only fantasize sometimes. But some traditions teach that *any* fantasizing should be avoided: that living a life in your head, either thinking about the future, the past, or fantasy, isn't a very good idea, regardless of whether those thoughts are pleasant or not. The idea of mindfulness is about paying attention to what's happening to you right now: sensations from your environment, as well as the thoughts and feelings that cross your mind. What mindfulness traditions often say is that you should not deliberately detach from the here and now and that attachment to the imagination is ill-advised.

Matthew Killingsworth ran a study using a smartphone app that asked people from time to time what they were doing, what they were thinking about, and how happy they were. It turns out that people mind-wander *a lot:* 46.9 percent of the time. Lovemaking seems to be the only activity that people really focus on. Supporting the Buddhist notion that you should be mindful of what you're doing, people were happier when they were paying attention to the task at hand, even if that task wasn't any fun.[30]

On the other hand, there is evidence to suggest that, at least sometimes, "living in the future" can actually increase your happiness.[31] For example, it appears that anticipation of a trip you're going to take (modestly) boosts happiness. Jeroen Nawijn found that people look forward to their trips, and this seems to make them happier before they go (this effect is very small: on a five-point scale, vacationers rated an average happiness of 2.25, and everybody else rated 2.07). This casts an interesting light on surprises. A surprise party can be fun, but it might be (for some people, at least), that a planned party increases happiness because they get to enjoy the anticipation of the party in addition to the experience of the party itself.

Compulsive fantasizing is an interesting case for mindfulness. On the one hand, the fantasies deliver real in-the-moment happiness. But too much of it can interfere with greater life goals. Many people have this kind of relationship to video games.

MEDITATION

Mindfulness *meditation* is deliberate, focused practice of being in the moment. All versions of meditation train you in controlling the contents of your conscious mind (whether the tradition emphasizes this outcome or not, this is a consequence of meditation). In a way, mindfulness meditation encourages you to *attend* to your imagination, but not to *engage* it. That is, if an image of something crosses your mind, you are to note that it's there, but you're not supposed to try to explore and elaborate on the idea, nor are you supposed to deliberately start using your imagination.

Mindfulness meditation is focused on reining in how most people think—the "monkey mind"—where they think about their bills, then something they should have said, how much it would hurt to give birth to a porcupine, then plan for something they need to do, and on and on. In mindfulness, you're supposed to just let your mind do whatever, and then allow the thoughts to pass without judgment. But with compulsive fantasizers, they have an incredible, often automatic, internal focus. How do you just let that happen and not judge it? The man I interviewed says, "I don't get tired daydreaming. It just . . . I can sit there, lie down daydream for a few hours, I wouldn't feel any less mentally exhausted from it."

I'd be interested to hear what a meditation teacher would make of a compulsive fantasizer, for whom thoughts don't just come and go, like the monkey mind of most people. Their thoughts, if left alone, will play out stories for hours, without any effort at all! On the one hand, they are simply attending to what comes to mind, which is what you're supposed to do with mindfulness meditation. But the way mindfulness meditation is often taught assumes that unrelated thoughts will come and go on their own, allowing one to focus on their breath or something else in physical world. But it's not like that for compulsive fantasizers, who might end up ignoring the external world entirely if they are simply mindful of their thoughts.

Mindfulness meditation is just one kind of meditation. Other kinds of meditation make heavy use of the imagination, particularly visual imagery. One might imagine being in a safe, happy place, or visualize different elements around your body, or imagine happiness bubbling up within you. Although some meditators might disagree with me, as far as I can tell visualization of this kind is not mindfulness meditation, because you're deliberately focusing on some mental construct that is not the here and now, rather than merely acknowledging what comes to mind automatically. Fantasizing, thinking about the future, or reminiscing are all things you're trying to avoid when you're being mindful. That's the whole point. But although mindfulness traditions emphasize being in the moment, in practice these traditions often utilize a combination of mindfulness and visualization exercises in their meditation instruction.

Interestingly, there is a downside to being in the moment—you are more likely to disconnect from your future self. The connection you feel to your future self predicts, in many ways, good behavior in the present: less procrastination, better eating habits, saving money, and so on. Recall that Eve-Marie Blouin-Hudon's experiment found that imagining your future self (which, again, is clearly not engaging in the here and now that's emphasized by mindfulness) resulted in a better connection to your future self than meditation did. The point is that whatever benefits mindfulness confers, it's not always the best mind-set to have. Sometimes using your imagination is good for you.

Mind-wandering isn't always a problem but can be when it proves a distraction from an important task. Fantasy can be pleasant, but like many things, too much of it can interfere with life goals. Excessive rumination on stressful and anxiety-producing thoughts are clearly harmful. So though it might not be a good idea to remove mind-wandering, in all its forms, it is probably a good idea to be able to stop it when you need to, and meditation is a training to help you do that.

8

Imagination as Mental Training, Healing, and Self-Improvement

We've established that when you imagine a scene, a good portion of your mind doesn't realize it's not real. That might seem alarming, but there is a bright side, too. Sometimes we can get the benefits of doing things without actually doing them, but by merely thinking about doing them. We can make changes to our bodies and minds just by doing things in our imagination.

In 1872, Alexander Bain was the first scientist to suppose that the body responded to imagery similarly to how it responded to the things we see, but the idea did not enjoy attention in the science world until 1931, when Edmund Jacobson did studies of mental imagery and related them to subtle muscle movements. More recent studies, using brain imaging, show that one of the main differences between actual and imagined movements is that in imagination there's another part of your mind that stops the message from actually going to your muscles—it's simply interrupted.[1] That is, in imagined movement, commands are sent to your body as usual, it's just that the message doesn't make it, like there's a glitch in your internal postal service. What this means is that, for the motor part of your brain, there is literally *no difference* between real and imagined motion, a little like a submarine captain who is doing all kinds of planning and issuing of commands that nobody ends up obeying.

Try it for yourself. Imagine putting your hands together behind you. Now imagine straightening your legs out, then spreading them. What you are engaging in, when you do exercises like this, is also a form of mental imagery and imagination. This kind of imagery is also very frequent in dreaming—you might dream that you're running, but your actual body's not moving. It's imagery. Part of this is visual imagery. You usually imagine yourself moving from a first-person point of view: your image looks like what it would look like if you were actually doing it. But a third-person point of view is also possible. In a third-person point of view, you see yourself as though you were watching a video of yourself doing the action.[2]

And although you might have a visual image motor movement, what's important here is what's called "motor imagery," which is the imagination of the movement and position of your body parts. When you're imagining doing something in third person, it's sometimes hard to distinguish the visual from the motor imagery (when it comes to therapy, one study found that third-person imagery only helps if it has a visual component[3]).

When you wake in the morning, even before you open your eyes, you know where all of your limbs are. This sense of your own body's position is proprioception (it's also sometimes called "kinesthesis"). Because it's a sense, just like hearing and vision, you can have imagery in that sense. That's what motor imagery is.

Just as visual imagery uses many of the same brain areas as visual perception, motor imagery tends to use the same brain areas as actually moving your body.[4]

When we engage in motor imagery, most of our minds think we're *actually doing* what we're imagining doing. Because so many of the same neural and bodily areas are activated during motor imagery as actual movement, your mind, and parts of your body, react the same way in both cases. For example, when people imagine writing, they experienced increased blood flow to those very brain areas activated during actual writing.[5] When you imagine exercising, your actual breath and heart rate increase.[6]

The exciting possibility is that mental practice might actually make you better at the real physical activity. This is indeed what experiments have shown. For example, in a virtual-reality based surgery training, doctors who mentally practice beforehand outperform those who don't.[7]

There is an enormous amount of evidence that simply imagining doing sports makes you better at doing the sport in real life. Mental practice, it turns out, is one of the few effective performance enhancing activities that exist at all.[8] One study, by Robert Woolfolk, had people simply imagine putting a golf ball into the hole just before they took their shot. The people who imagined doing it correctly were 30.4 percent better than those who did not![9]

As you might imagine, mental practice is most beneficial for those activities with a strong mental component.[10] Basketball, for example, involves searching, planning, coordinating with other people, etc. So mental practice is more beneficial for these complex activities than they are for simpler things, such as simply applying a force, as in weight-lifting. But that is not to say that imagination does not help for simpler tasks—perhaps the most shocking finding is from exercise scientist Guang Yue, who showed that imagination can even be used to increase muscle strength! Dr. Yue found that finger strength could be increased by 22 percent, which is almost as much as doing isometric exercise, which increased strength by only 30 percent.[11] There is also some preliminary evidence that sports practice during lucid dreaming can improve performance in the real world as well.[12]

So should you stop actually practicing the sport and spend all your time imagining doing in from the comfort of your sofa? No. It turns out that twenty minutes is the optimal amount of time for a mental practice session. Any less and the virtual practice doesn't stick as well, but if you do more than that you can start to lose touch with reality and interaction with the physical world, practicing things that simply won't work in real life.[13]

This is a view of imagination as a kind of virtual practice. And just as imaginary practice can help you, so can virtual practice using computer-generated imagery. Just as much of the mind can't tell the difference between what's real and what's imagined, it also can't tell the difference between what it sees on a computer screen and what it sees in reality. Our conscious, deliberate minds know that it's just a screen, but we quickly get lost in the virtual reality of what is presented on it, be it a movie, training simulation, or video game. And "training" simply by playing video games transfers to skills in real life. Certain video games, for certain people, can lead to depression, addiction, and aggression. However, games can also improve attention allocation and improve visual and spatial processing. Interestingly, the games most likely to cause aggression, violent "shooter" games, are the ones most effective at improving these skills—comparable to courses in school specifically tailored to improving these skills.[14]

The virtual reality of our imagination can be put to other uses, too, including dieting. You might think that fantasizing about eating food will make you hungry to eat more of it. But a study by psychologist Carey Morewedge showed that simply imagining eating a food, like cheese, makes you habituate to it, and consequently you eat less cheese in real life. It appears that imagining eating food makes part of your mind, at least, think you actually have eaten it, and you feel more satiated.[15] You have to be careful with this, though. Merely thinking about food, without imagining eating it, probably makes things worse. A review by Charles Spence showed that looking at pictures of food, such as Instagram, makes people hungry and salivate, wanting to eat.[16] To be satiated, you have to vividly imagine actually doing the eating!

MNEMONICS

So imagination can be used to make you better at sports and can help you diet. Can imagination be used to help your memory?

Visual imagery, if you have it, has been shown to be very helpful for use in remembering. Specifically, if you have to remember a list, using imagery in the right way can help immensely. A mental memory aid, or mnemonic, is the use of something easy to remember to help you with something that is hard to remember.

What kinds of things are easy to remember? Well, concrete concepts are easier to remember than abstract concepts. For example, pigs and bricks are concrete, and some things are inherently abstract, such as justice, the number 287, and how some people have a penchant for mentioning pensions.[17] It's also easier to remember things that are important for survival and reproduction—particularly if those things existed when our ancestors were trying to eke out an existence on the plains of ancient Africa.[18] Dangerous, violent, sexual, and emotionally charged imagery, in general, is easier to remember. A study by Rolf Zwann had people read two versions of a text. In one, there was a pushpin left on the floor where somebody might step on it. In the other, the pushpin was safely stowed in a box. People were better at remembering the pushpin on the floor, suggesting that objects that are likely to be relevant to future events are more easily remembered.[19]

So what's hard to remember? Numbers are notoriously difficult for most people to remember. One technique created to help people remember strings of numbers (such as your credit card number) is the major system. You can use it to remember numbers by recalling a word or phrase—because words are easier, in general, to remember than numbers.

In the major system, you associate different consonant sounds with each digit. 0 is the z or s sound, 1 is the t, th, or d sound, 2 is n, 3 is m, 4 is r, 5 is l, 6 is ch, sh, j or a soft g, 7 is k or a hard g, 8 is f or v, and 9 is p or b. Importantly, there are no digits for vowels.[20]

So you can take any word or phrase and get a list of numbers from it by paying attention to the consonant sounds in it (it's the sound that matters, not the spelling.) So, for example, the phrase "I play the acoustic Theremin" encodes the numbers 95170171432. Sometimes, when I'm walking along the street, I practice the major system by turning street names into number sequences.

To use it to remember numbers that are important to you, you create a phrase that encodes them. Suppose, for example, somebody tells you that the entry code for their door is 4125. You can't write it down, and you're busy, so you can't just rehearse the numbers over and over until you need it. You can try to make a word or phrase from it. Let's say it will have r, d, n, l. I look at those letters and see the phrase "redden Leo." Then, when I get to the door, if I can remember "redden Leo" I can decode it into the numbers 4125.

If you care to, you can do this with your spouse's government ID number, important phone numbers, credit cards, and whatever else. It works because phrases are easier to remember than long numbers.

MEMORY PALACES AND METHOD OF LOCI

But perhaps the most important and popular mnemonics are memory palaces. In a memory palace, you imagine a walk-through of an environment you know well, such as your childhood home. You mentally place objects at different locations along the route, and later you can do a virtual walkthrough of the environment and see, in your mind's eye, what you put in those locations.[21]

I'll explain with an example from my own life that I use often. I ride a bicycle to work. It takes me about half an hour or so, and I often think of things on the way that I need to remember. If it's important enough, I'll stop the bike and make a note on an index card (I keep a stack of these in my pocket at all times). But I am not always eager to do this. I live in Ottawa, where the winters are so cold that even the disco balls retract. So if it's forty below zero (nobody should have personal experience that this temperature is the same in Fahrenheit and Celsius), writing it down would require stopping, removing my mittens, pulling the cards out from beneath my snow pants, and generally taking a lot of time and getting really cold. Here's where my memory palace really comes in handy.

I have an imaginary walkthrough of my childhood home. It starts in my bedroom. I'm sitting on my bed, and I look to the left and see my old turntable. That's the first location. The desk is next to that, so I look at what's on the desk. Then I look out the window at what's beneath the tree in my backyard. And so on. I have over seventy locations in there. Remember, this is all in my imagination—I haven't actually been inside this house since the 1980s.

Suppose I want to remember to call the dentist. I'll picture my dentist on the turntable. To make the image more interesting I might make her run on the turntable like it's a treadmill. Later on in my ride, I might remember that I want to have a more Chinese cultural influence in the novel I'm writing. I'll picture an image of an ancient Chinese building on my desk, shrunk down as though it's a miniature. To remember that I want to associate it with my novel, I might picture a book above it, open like an umbrella, shielding the building like a giant, floating umbrella.

Sometimes, by the time I get to work, I have seven or so things in the memory palace. At work I do a walkthrough and jot down the ideas. Dentist, using Chinese imagery in my novel, and so on. Then I "clear out" the memory palace by imagining each location and removing the objects I've placed there—basically imagining the locations empty. It's really amazing how well this works. Studies show that using this strategy helps people remember two to seven times more things than people who don't.[22]

If you're going to try this yourself, here are some important things to keep in mind. Make sure it's a place you know well. Human beings are very good at remembering routes through environments, and what objects are in what locations, which is why the system works so well. But if you forget what's where, the system won't work.

The walkthrough should be canonical—that is, you should use the same walkthrough order every time you use it, and the locations in the palace need to be attended to in the same order every time. For example, I will never make the first thing I need to remember on the desk. It's *always* the turntable first. This helps you remember things in order, which is normally quite hard. It also ensures you attend to everything in the palace. You don't want to have to go to every location to make sure you recalled everything. When I am trying to retrieve things in the walkthrough, I go through it until I see a spot with nothing there, and then I know I'm done.

You also need canonical locations. In my actual childhood room, there is a bit of floor between the turntable and the desk, but I don't ever use that location to remember things because I've chosen some places, and not others, to be "the locations." You can rehearse going through the locations even without having anything in them. I often do this to help me sleep at night. I close my eyes and imagine walking through my old house, looking at and noting the importance of each location, even if there's nothing currently in them, just to practice getting them all in order.

When placing the objects, they should be memorable. As I mentioned before, concrete things are easier to remember than abstract things. So using concrete symbols of abstract concepts is useful. Suppose your friend Liz recommended that you listen to a particular song, and while riding your bike you want to remember to tell her that you listened to it. Imagining telling Liz that you heard a song is pretty abstract. Telling someone can happen in many ways. An MP3 doesn't look like anything. The only concrete thing about the scenario is Liz. So I might picture myself talking to Liz using one of those old tin cans connected by a string telephones. The picture of myself will have me dancing in front of an iPod. This will probably be enough to recall what I meant later. Of course, sometimes I misunderstand my images. Later, I might think to myself, "Why did I want to talk about dancing to Liz?" You won't always be able to correctly decode what you put in the palace, but the silver lining is that you can be amused by your own incomprehensible imagery.

I've heard people advise imagining bizarre things, and when bizarre things are in lists with normal things, the lists are better remembered. But in memory palaces, the evidence is a bit mixed. Bizarre things are not better recalled, in general, but they might be less subject to interference with other memories.[23] And back to our species' survivalist days, violent or sexual imagery works best for me. I'll spare you an example.

The memory palace technique has worked so well for me that I've made myself several of them. The one of my childhood home is like my working memory—nothing is permanently put there. It's simply there for temporary storage like remembering things I think of on my bike ride to write down when get to work—then it gets regularly emptied.

That's my temporary memory palace. I also have a few permanent ones. These are palaces that I maintain and might add to, but never clear out. They are for things I want to remember for the rest of my life.

One permanent memory palace I have based on my parent's island cabin, which I use to remember the lyrics to my favorite album: Paul Simon's *Graceland.* The first location is the little motorboat we use to get from the mainland to the island. The first track of the album is called "The Boy in the Bubble," so I picture a kid in a large, plastic bubble taking up most of the space in the motor boat. Walking on the water, on the way to the island, I picture soldiers walking slowly, with the sun shining on them, which I can decode into the first lyric of the song. It goes on like that for the whole album.

Now, because I know the album so well, and with help from the memory palace for track order and lyrics, I can recall the entire album in my imagination. This is great for when I'm riding my bike (I can sing most of the album on my way to work), when I am loathe to wear headphones because I want to be able to hear the traffic. One time I was on a long drive, and my wife fell asleep. I didn't want to wake her with music, so I just "listened" to *Graceland*, completely in my imagination, and managed to entertain myself with it for about forty-five minutes. Diana Nyad is famous for swimming from Cuba to Florida without a shark cage. She has about 120 songs in her head that she "listens" to for long-distance swims. Swimming, you can't have headphones, so you have to entertain yourself in your own mind. I don't know if she uses any particular memory strategy to remember these songs.[24]

Another memory palace I'm working on is for remembering dates. I'm bad at remembering numbers of any kind, so I use mnemonics to remember the numbers I really want to. In particular, there are dates I need to remember. So I used the entire geography of North and South America as a memory palace.

It starts in a large, upstate New York house my parents once lived in. I made a memory palace for it, with a location for each year I've been alive, and each year I expect to be alive (I used an online estimator to get an expected year of my death). So I have a location for every year of my life, past and future.

Then I have a different location in my hometown for every decade, going back a few hundred years. Across the street is a YMCA. I use this to represent the 1960s. The Friendly's restaurant represents the 1950s, and so on. Within the YMCA, I have a different location for each year of the decade (1960 through 1969). Similarly, I have locations for each year of the 1950s in the Friendly's. This goes on until about 1500, when I get to the end of my hometown, traveling along the main thoroughfare. From there, I travel south, basically following the east coast of the Americas, all the way to the South Pole. As I travel south, I mark cities with dates. New York city represents 1,000 years ago, Philadelphia 2,500 years ago, and Washington, DC, 5,000 years ago. The time spans get greater and greater as I go south, ending with Antarctica, which I use to represent the big bang, estimated to be about 13.4 billion years ago.

This one is a bit more difficult because I have to memorize three things: the list of ordered locations as they appear on a map, what year they are associated with, and then what objects I choose to put there, representing what happened around that date. For example, the last common grandmother of both humans and chimpanzees lived about five million years ago, so I have an image of some missing link human–ape creature watching the sunset in Key West. To make this work, I have to associate Key West, somehow, with the creature and the number five million. This is difficult, and I'm still working on it—I don't yet have it all memorized.

This can be further complicated by the fact that I sometimes want to put more than one thing at a location. More than one thing happened in 1986. Just in the world of hip-hop, two classic albums came out that year: *Licensed to Ill* by Beastie Boys and *Raising Hell* by Run-DMC. It was also the year of the Chernobyl nuclear disaster. Recalling all of these things is easier if the images are integrated. You might imagine Run-DMC and Beastie Boys partying and getting sick around a nuclear cooling tower. As I add more items to the location, I should strive to combine the images into a coherent scene. Otherwise I'm just trying to remember a list, which is the difficulty memory palaces are supposed to overcome in the first place.

Note that you can combine mnemonic systems: I can put images that generate phrases that trigger major system number lists.

I also have a memory palace to remember my interests (in case I ever want to list them), for months of the year, my favorite jokes, and for card manipulation (cardistry) tricks to practice. I have enough memory palaces now that I created a memory palace to help remember all of my memory palaces. For that, I used a geography that's a bit different: the Fisher-Price toy castle I played with as a child. Even a small thing like that has distinct locations that I can remember in order and do a virtual walkthrough of. The wonderful thing about imagination is that I can shrink myself down to a size such that I can walk through this toy castle as though it were a real place. I've heard of people using the environments of 3-D shooter video games they know really well as memory palaces. If you know the environment well enough, they can be used in this way. When I'm trying to sleep at night, I choose which memory palace to review based on this castle memory palace (the first location is Monday, the second is Tuesday, and so on). This way I focus on a different palace each night of the week, and I get to rehearse all of my memory palaces roughly the same amount.

But before you think I'm completely nuts, I'm going to stop talking about my memory palaces and move on to something else.

HOME ECONOMICS

Time to use your imagination. Imagine that you are about to buy a milkshake, and it's $10. Then you hear that there's an identical milkshake for sale. It's five blocks away, and it's only $5. Would you walk the five blocks to buy the cheaper milkshake? Most people would.

Okay, now imagine you're about to buy a smart watch for $505, and you hear that there's an identical one for sale, five blocks away, for only $500. Would you walk the five blocks to buy the cheaper smart watch?

If you're like most people, you'd be more eager to walk five blocks to save $5 for the milkshake than you would be for the smart watch. And why? Well, it seems like $5 is nothing compared to the $505 purchase you're considering, but it feels like a big discount for a $10 shake. But this is really irrational—the effort you're putting in, walking five blocks, is the same in both cases. In both cases, you're basically getting a value of $5 for walking five blocks. The actual amount saved is independent of the overall cost of the product you're buying.

You can try to overcome these kind of irrationalities by using your imagination. In this example, you might try to abstract away from the purchase you're making. Try to forget whether it's a $10 shake or a $505 smart watch, and just think about the monetary value of the added effort. Would you walk five blocks for $5? What if some stranger offered you $5 to walk a five-block round trip to drop off something? Would you take this gig?

If you walk a block a minute (which is pretty normal), then this is getting paid $1 per minute. How long would you walk for $1 per minute? Would you do it for two hours ($60 per hour)? The answers to these questions are probably pretty clear in your imagination, and it helps when you abstract the issues like this.

How much is your time worth? If $60 hour is more than you make at your job, perhaps it's worth walking five blocks in either case. If you make $300 per hour, maybe not. Of course, this also depends on how much you'd benefit from a walk, independent of the money. Is it raining? Do you need the exercise? Or a break?

My beloved and I used this reasoning when we hired someone to clean our house. We were doing the house cleaning ourselves, taking turns. I would do it one week; she the other. It took us a few hours every time. We found that a cleaning person would do it for $90.

My beloved asked me: would you clean our neighbor's house every other week for $90? My answer was a clear "no." She asked me how much money I'd charge to do this. The thought of cleaning my neighbor's house sounds so annoying that I admitted that I'd probably charge $500 per hour. Otherwise it just wouldn't be worth it to me. Well, the answer was clear—it wasn't worth my time to clean my own house either. We hired a cleaner, who, it turns out, does a better job at it than either of us ever did.

When I was in graduate school I didn't have as much money. It was also in the era before e-tickets for airplanes. I'd buy my ticket through a travel agency. They'd get the ticket, and I could either go pick it up, or they would mail it to me for $10. After using this same kind of reasoning, I had them mail it to me. Getting there to pick it up was onerous. Parking was difficult, and the whole enterprise probably took forty-five minutes.

My friend was still picking his tickets up. You know, to save $10. I asked him: "I'd rather give the money to you than to the travel agency. How about, the next time I need a ticket from them, I pay *you* $10 to go get it for me? Would that be worth it to you?"

Even though I was a friend, his answer was a clear no. His time, even as a graduate student, was worth way more than that to do something so boring. So why do it for yourself?

A little while ago we went to Disney World. They had more expensive tickets that would give you FastPass and shorter wait times in lines. I was trying to save money, so I got the cheaper tickets. That was a mistake. I didn't apply my imagination correctly! How much would I have to be paid to wait in line for somebody else? A hell of a lot more than the FastPass tickets cost, I can tell you that. I'll know better for next time.

It's not that using your imagination in this way can save you money, but it can help you get better value from how you spend it.

Your imagination can also be used to improve your health.

HEALING AND PAIN RELIEF

Using your imagination can help you heal faster. Using guided imagery, in addition to traditional physical therapy, resulted in greater improvements in the stability of the knee for people who had undergone surgery to their ACL, a ligament in the knee.[25] Stroke survivors with hemiparesis were given motor imagery tasks to do and showed improved hand and leg performance as a result.[26] Other studies have shown similar effects, but it's difficult to know whether it was the imagery itself that did the work, or the fact that imagery caused stress levels to decline.

Using imagery, particularly when guided by a therapist, has been found in many studies to be an effective form of pain treatment—even better than distraction, self-talk, relaxation, and biofeedback. For example, in studies where participants put their hands in ice water, using imagery tends to reduce feelings of pain. But the effect on pain *tolerance* is even greater. When people are asked to put their hand in ice water for as long as they can, those using imagery enjoyed subjective feelings of pain reduction by 30–40 percent, and were able to keep their hand for double the time of the nonimagery group.[27] In spite of its effectiveness, imagery is one of the most underused pain treatments.[28]

The studies that produced the impressive results I'm reporting here were done under strict experimental controls. There are drawbacks to experimental control, usually, and one of the drawbacks here is that the practicality of running experiments can prevent long sessions of imagery. The strong results that reduced feelings of pain by up to 40 percent were mostly done with studies that involved only about two minutes of imagery![29] In clinical sessions, imagery is done for much longer and repeated over time. So these results might be actually *underestimating* imagery's effectiveness, if longer imagery sessions are more effective.

It's tempting to think that this is mostly a placebo effect—that is, when a treatment works simply because people believe it will work, not because of the efficacy of the particular intervention. It is difficult to come up with a placebo condition in an imagery experiment—in fact, it's hard to enforce a nonimagery condition, because people so often engage in imagery even if they're told not to! In spite of these issues, studies show that the effectiveness of imagery on pain reduction has nothing to do with whether you even believe it will work, casting doubt on the idea that the placebo effect is a big factor in imagery treatment.[30]

IMAGERY TREATMENTS

In a therapeutic setting, imagery is often guided by the therapist, who uses some kind of script, often describing some relaxing scene on the beach. When these imagery scripts are written down, there is a tendency to treat them as a carefully worded transcript that should be followed to the letter. However, it's best for the pain sufferer to help modify the imagery or come up with it on their own. A beach scene might not be best for somebody who was on the beach when their spouse told them it was over.

Some guided imagery scripts are rather vague so that the imaginer can flesh out the content themselves. For example, you might say to someone, "Imagine that you are going to your own special place. It is safe and makes you feel relaxed, content, and happy. It might be a place you would like to go to, or could be completely make-believe. Observe it carefully. What are the colors, the smells? Where are you in the space?" And so on. In general, these relaxing scenes should use multisensory imagery—temperature, textures, sounds, and smells—in addition to visual imagery.

But sometimes people need a little more structure to their imagery. Sometimes cueing with specific images can be more powerful than what they'd come up with when you ask them to imagine some special place on their own. No matter how much detail you give them, they're still going to have to make a lot of it up. Some people are good at this and find it easy, and others can't do it or need to put in a lot of effort.

There are some things that work better than others, in general. In particular, researchers and practitioners have found that beaches, mountains, mountain cabins, and gardens have broad appeal.[31] The use of light imagery is very commonly used and is also used in many traditional shamanist practices to represent healing, cleansing, and rejuvenation. Treatments might ask the patient to imagine filling herself with a radiant light.[32] The ubiquity of light imagery being associated with health is because we associate light with goodness and darkness with evil cross-culturally.[33]

IMAGINED LIGHT

When I was speaking at a conference I met Jennifer Veitch, a psychologist who studied the effects that seeing light had on the mind. Bright light, especially blue light, keeps you awake, which is why it's not a good idea to spend much time on your cell phone before bed.

Light can also make you happier, and sometimes therapists use light therapy for depressed clients. Knowing that imagery can often be used in place of reality, I asked her if she thought that you can get the benefits of seeing light from just imagining it.

I was delighted to hear that a study had actually been done! It turns out that you can't get the benefits of light from just imagining it. Brenda Byrne had people imagine seeing very bright light, and then measured their melatonin levels, which also react to seeing actual light. Turns out it doesn't work. Just imagining light might improve your mood, but it doesn't affect your physiology the same way that real light does, nor does it make you less sleepy.[34]

Simply imagining oneself in special places causes relaxation, which itself causes pain reduction, partially because people are simply distracted by something pleasant. The emotion caused by the imagery is positive, counteracting the negative emotion associated with pain (happiness versus unpleasantness).[35]

Also, people can think of their pain in a variety of metaphorical ways—a burning, or as needles, or aching. "Fire" pain might be best treated with water imagery, and "needle" pain might be best treated by having the patient modify the needles in their imagination (such as making them so wide they no longer hurt).[36] This is an example of using incompatible sensory imagery. If a person is feeling pain because of cold, for example, she might benefit by imagining a hot day in the desert.[37]

Some therapists and patients picture the pain as some animal, such as a tiger biting the spine. The patient can then work with the imagined creature, befriending or taming it, to help relieve the pain.[38] The transformation should be appropriate for the image: taming a cat would involve different imagery than putting out a fire. These metaphors can be incorporated into the imagery therapy.[39]

Pain can also be reduced using a technique known as "sensory transformation," which works by focusing on the pain and then transforming it in the imagination into something more pleasant. For example, one might take the sharp pain of a burn and try to transform it into an itching, numbness, or tingling. Interestingly, trying to transform it to something *too* pleasant is so incompatible with pain that it doesn't work as well. This kind of thing does not work instantly, so the transformation must be gradual. In fact, the therapist can take advantage of this time delay when guiding the meditation by saying something like "you won't feel any change at first." This gives the patient the impression that even though nothing yet is happening, it is to be expected and the therapist knows what she is doing.[40]

Another interesting treatment is called "glove anesthesia" in which the patient uses imagery to feel that their hand is numb. After that, they are guided to believe that they can transfer the numbness from their hand to another part of their body with a touch. Then, they rub the painful body parts to transfer the numbness. This is especially good for localized pain that flairs up occasionally (such as lower back pain and arthritis).[41]

Displacement attempts to gradually reduce pain by slowly changing an image associated with it. Displacement guides the patient to move the pain from one location to another in their mind. With practice, the image representing the pain can be moved outside the body, reducing the felt intensity of the pain. Dissociation is related but can involve imagining that the painful part of your body does not exist (for example, imagining that you have no arm) or that the patient has left their own body and are looking at it from some distance. These methods are supposed to allow the patient to be aware of the pain, but suffer from it less.[42]

Other direct reduction scripts involve numbers. For example, you ask the patient to imagine their pain as the number seven, and to picture that number clearly and to change it into a six. With practice, the patient gets the number lower and lower, and the pain reduction comes with it. This can also be done by imagining a dial for the "pain gate" in the brain that lets the pain information through.[43]

Further interventions reduce pain by changing the *meaning* of the pain in the imagination of the patient (called "contextual transformation"). Arm pain has been treated by having the patient imagine that she were a spy who'd been shot in the arm and was escaping from enemy agents! Giving the pain some kind of (imagined) purpose made it more tolerable or, perhaps, even a point of pride. This has also been done with the pain of childbirth, transforming it from a more sterile, clinical pain to a rite of passage of motherhood, represented as a journey to meet the new baby.[44]

Pain management is a huge problem in medicine, and, as I have shown, there are many imagination methods created to deal with them. Unfortunately they are underused in practice, despite having a real chance of positive results and little downside or side effects. Imagination has also been applied to other mental health issues, such as anxiety. The step-up technique has been used to treat people with fears about the future. It involves the therapist asking the patient about the consequences of the terrible things they imagine happening. Doing so often results in the patient realizing that their fears are rather far-fetched, and they become less anxious.[45]

MEDITATION AND CREATING EMOTION

Meditation is one of the most common therapeutic practices to use imagery. I've been meditating for many years now. At one point in my life, for a year-and-a-half, I meditated for about twenty-five minutes a day. I was happier, and better able to deal with the annoyances and hardships of daily life better, but I could not tell if it was from the meditation itself or the books I read to learn how to do meditation. These books gave me great new perspectives on how to manage my own mind and feelings.[46]

For several years I meditated about once per week with a group consisting, coincidentally, of three lawyers, one law student, and three computer scientists. I told my meditation teacher, Patrick, about how I was trying to make my visualization more vivid. One of my goals for this was to allow myself to mentally put myself into a "happy place" (for me, a screened-in porch by a river) to make me happier when I was feeling down. I didn't want to become a compulsive fantasizer, but I wanted to be a *little* more like that, able to entertain myself with my own imagination more fully.

What I was doing was generating an image, the content of which made me happier. It was a kind of fantasizing. Patrick suggested that using imagery was the first step, but ultimately that was making myself happy through distraction. Much like watching television, but in your own head. (Not that this is any small benefit! It's great being able to distract yourself without external media.)

In the long term, though, you don't want to use imagery as a crutch. He said that the next level was to be able to generate happiness all by itself. That is, rather than willing imagery to mind that I get happy about, simply willing the happiness directly, without the middleman of imagination. That is, instead of creating a picture, just create the happiness. This idea kind of blew my mind.

With practice I've gotten better at generating happiness directly, using less and less imagery. I'm still trying to improve the vividness of my imagery (which might not be possible), but I'm also working on being able to be happy without requiring anything in particular, real or imagined, to be happy *about*.

You might be skeptical of this. Is it even possible to just will happiness like you might will yourself to wiggle your toes, or to imagine opening a jar of peanut butter? I can only tell you that it seems to be working for me. (I have been unable to find scientific studies on this.) Skeptics might well ask if it's all in my head. I would counter that, yeah, of course it's all in my head. It's happiness. Where else would it be?

CAN YOU BE WRONG ABOUT YOUR OWN EMOTIONS?

There seems to be a close tie with our emotions, such as happiness, and our conscious feelings of them. For the most part, if you think you're happy, you probably are. But is it possible to be depressed and not know it? Or to be happy and not know it? Again and again we've seen that although we're conscious of lots of things going on in our minds, we're often oblivious to many others. Can you have subconscious emotions, or, as I like to ask, can you have emotions you can't feel?[47]

Several brilliant scholars of the past thought not, including William James and Sigmund Freud. But there are some reasons to think that unconscious emotions happen.

Donald Dutton ran an ingenious experiment. He had one group of men cross a scary suspension bridge, and another group of men cross a bridge that felt perfectly safe. Halfway across the bridges, the men would be approached by an attractive female interviewer, who asked them to fill out a questionnaire. The men on the scary bridge had more sexual content in their answers and were more likely to try to contact the comely interviewer after the experiment. Not so much for the men on the safe bridge. What's going on here?

It turns out that sexual attraction and fear are accompanied by some similar physiological responses, such as increased breathing and heart rate. But what was happening here was that the men were (subconsciously) interpreting what was happening to their bodies as sexual attraction, when actually the physiological changes were, at least in part, driven by fear. It was a misattribution.[48]

Helen Fisher reports that there was once a male graduate student who knew about this effect and wanted to use it to make another student attracted to him. During a conference in Beijing, he invited her to go on an exciting rickshaw ride around the city. She squealed and clung to him during the ride. At the end, she said "Wasn't that wonderful? And wasn't that rickshaw driver handsome!"[49]

So we can see that people can have one emotion and mistake it for another, or, at the very least, that physiological changes result in different emotions, depending on the context. And we can be wrong about what causes our emotions. That's not exactly the same thing as having an unconscious emotion, though.

One thing we know about emotions is that the subjective feeling of an emotion, such as anger, tends to accompany changes in the body, such as sweaty palms. But these don't always go together. Therapists who deal with people with anger issues will sometimes recommend that they pay more attention to their own bodies, so that they know when they are getting angry sooner. Like many basic emotions, anger is accompanied by several physiological changes, including sweating palms and clenching of the jaw. By learning to pay more attention to their own bodies, the clients learn to be able to notice they're getting angry earlier, so they can do some self-care before they blow up and do something they'll regret.

But notice that the whole premise of this idea is that someone might start to get angry but *not know it*. Essentially, this means the person is having the emotion but are not (yet) aware of it. They are focused on *what's making* them angry, and this focus eats up all the attention, so that they have none left over to monitor their own emotional state!

Jocelyn Sze ran a study in which people were shown emotional films. They wanted to see how much their changes in bodily measures, such as heart rate, correlated with subjective experience of emotion. For example, when watching a scene of one man crushing another man's head, they measured both changes in heart rate and how positive or negative the person was feeling (the study participants used a dial to indicate how good or bad they were feeling moment-to-moment). The idea is that the body might be in a negative emotional state that the conscious mind might not be completely aware of. She looked at three groups of people: experienced Vipassana meditators, dancers, and a group of people who were neither. Vipassana meditation focuses on bodily sensations, such as heart rate and breathing. She found that people who were more in touch with their bodies had more coherence between their bodies and felt emotions. In terms of the heart rate lining up with felt emotion, the meditators were the best, the dancers second, and the control group third.[50]

These studies make sense of what therapists are trying to get their clients to do: pay more attention to their bodies so they can better know what emotions they're having!

Perhaps the most convincing study was done by Piotr Winkielman. He had people look at happy and sad faces, and then taste a lemon-lime drink and rate how good it tasted. But the faces were displayed on a computer screen so quickly that the research participants had no conscious awareness that a face was even presented, let alone know whether it was happy or sad. It was subliminal imagery. But the people who were presented with happy faces rated the drink as tasting better and drank more of it! It seems that subliminal imagery can affect people's emotional states without their knowing it—they were not consciously aware of any mood change. But they must have felt better, at some level, because they liked the drink more. The authors speculate that our ability to consciously feel our emotions might have come later in our evolutionary history than having unconscious emotions, suggesting that unconscious emotions are actually more fundamental and primal than conscious experiences of them.[51] Subliminal and conscious emotions also have different brain signatures.[52]

If our emotions evolved to help us react appropriately, consciousness of them might not have been a requirement.

Whether it's conscious or not, imagery and imagination are famously important for creativity. Imagination and imagery are clearly important for design tasks, such as in architectural, graphic, and industrial design. That's kind of obvious, because they are essentially spatial and visual domains. But visualization has been shown to be important even for things like computer programming.[53]

Although we often talk about creativity and imagination as being the same thing, I'd like to distinguish them a bit. To help us do this, we can look at how some artistic creation involves more imagination than others.

At one extreme of imagination we have people who do enormous amounts of creative work in their heads before making anything in the real world. Rapper Kanye West claims to not have written down any of the lyrics for his first four albums.[54] This means that he composed all of the lyrics in his imagination. At the other extreme are rappers who freestyle, rapping off the cuff and making it up as they go along. The difference here is in the amount of planning. If we think of imagination as being the construction and reconstruction of ideas in the mind, composing four albums requires far more imagination than rapping off the cuff, where we only have the cognitive capacity to plan ahead a few lines, or, in some cases, a few words. It's just happening too fast. They are both intensely *creative* tasks, but they differ in the amount of *imagination* used.

Rapping off the cuff is a kind of improvisation, which can happen in many art forms. Musical improvisation, most famously done in jazz, often happens with a whole band. You can't plan ahead very far because you don't know what the others are going to do. It doesn't make sense to have an agenda in your head, because by the time you get to those future planned points in time, the context will be completely different from what you expected, and what you'd previously planned will no longer work. With all kinds of improvisation, it's better to listen carefully and stay in the moment, doing what's needed at the time.

I did theatrical improvisation for about twenty years. I was onstage with other actors, making up scenes as we went along. In my improv training, I was explicitly told not to have an agenda, because it would interfere with the scene quality by making me less responsive to what was happening on stage. Getting better at improv involved enormous practice so that I could simultaneously pay attention to staying in character and talking, while at the same time looking critically at the scene and deciding what the story needed—an emotional change, a reincorporation of a previous idea, things like that.

We can see improvisation, or something like it, in graphic arts. Salvador Dali made hundreds of sketches, changing and manipulating as he went, before he made a final painting, but other artists are more like freestyling rappers. When I do gestural calligraphy, where you make large motions with your arm, instead of fine motions with your wrist and fingers, I often have a vague idea, or no idea at all, of what's going to be on the page before I start making it. Making art is about being responsive to what's been done before.

In saying that improvisation uses less imagination than more planned art, I want to emphasize that I'm not saying that improvisational art is less creative. Both kinds of art creation are intensely creative. It's just that one involves more preliminary mental construction than the other, and it is this mental construction that is connected to imagination.

CREATIVITY

The term "imagination" is often used to be synonymous with "creativity," so it should come as no surprise that using your imagination can enhance your creative abilities. Nicholas LeBoutillier did a review of studies and found a mild but real association of imagery ability and creativity.[55]

Most people think of imagination as imagining something creative—creating, in your mind, a world that does not exist. We can think of realms of fantasy and science fiction or imagine flying in our dreams. The "imagined worlds" I've been talking about so far in this book aren't particularly creative. What about the fantastic flights of creative imagination we see in the creation of fantasy and science fiction worlds that authors and filmmakers create, like the worlds of Star Wars or Harry Potter?

What's interesting is that when you look closely at these worlds, they actually look a whole lot like the real world. There's gravity, solid objects generally can't pass through one another, there are trees and wind. Also there are people, who speak with language, reproduce, bleed when cut, have dreams and preferences, and would probably think that drinking grapefruit juice right after brushing your teeth would be disgusting. The fantasy and make-believe elements are actually a small part of these worlds. The audience needs a lot of things they recognize to appreciate the details that are unusual.

It takes a good grip on reality to make compelling fantasy.

We use our imaginations not only for creating other worlds, but for creating alternate histories in our own lives. As mentioned in the chapter on memory, we use our imagination when we visualize things that have happened to us in the past or imagine things that could happen to us in the future. But sometimes we "remember" what might have been. We think about something that could have realistically happened, but didn't—for example, you might think about what how your life might have changed if you'd attended a different university. This counterfactual thinking is importantly different from pure fantasy. With counterfactual thinking, we use what is called "reality monitoring," which is the use of our knowledge of how the world works to help keep the imagining realistic.[56]

For example, you might imagine having gone to Harvard University rather than Reed College, but you would be unlikely to imagine that because of this you would lose four inches of height, or grow wings, because you know those things are impossible. Counterfactual thoughts do not particularly feel like the products of creative imagination; they are realistic alternate histories that adhere to our understanding of how the world works. If you describe counterfactual scenarios to people, they will reject them as implausible if they stray too far from reality.[57]

So when do we get creative?

CREATIVE PROBLEM SOLVING

One important function of creativity is in problem solving. There is a famous experiment in psychology called the "Duncker candle problem." Here is how the problem is often presented—try to solve it yourself before reading on.

Your task is to affix the candle to a cardboard wall so that the candle burns properly and does not drip wax on the table or floor. You are given a candle, a book of matches, and a box of tacks.

What most people do is try to use melted wax as an adhesive or to tack the candle to the wall. Neither method works. Only about 23 percent of people correctly find the solution to this problem.[58]

The most common stumbling block is something called "functional fixedness." People focus on the usual function of the box: as a container for tacks. They fail to realize that the box can be used as a stand for the candle. The correct solution is to tack the box to the wall as a stand for the candle—the matches aren't used at all.

Psychologists have found numerous interventions that increase the rate of solution finding, but I'm going to talk about the ones relevant to imagination. Adam Galinsky found that one could increase peoples' chance of finding the solution by getting them into a "counterfactual mindset" that encouraged mental simulation. To do this, the experimenters had people read scenarios about things that almost happened or didn't. For example, in one, someone *almost* won a trip to Hawaii at a rock concert, but didn't because her seat got switched. When people think about situations like this, it puts them in a mindset of thinking about what might have been, which seems to increase mental simulation. Putting people in this mindset brought the solution rate for the candle problem from 6 percent to 56 percent, effectively increasing their problem-solving creativity.[59] That's a big jump. The creativity lesson here is the next time you are stuck on a problem, try first imagining how things in your own life might have been different.

It gets more complicated. It turns out that *how* you imagine makes a difference. To understand the science behind it, it's important to understand the difference between additive and subtractive imagination. If you are thinking of the world much as it is but *removing* an element that would be true in the world, it is called "subtractive." For example, if you try to imagine what your day would have been like if you hadn't been reading such a great book, you are engaged in subtractive imagination. If, on the other hand, you imagine what your day would have been like if you'd taken a walk after breakfast, you are engaging in what's called "additive" imagination. Although there are some instances of imagination that are hard to classify, independent raters have a high level of agreement regarding whether an imagining is additive or subtractive.

Additive imagination encourages a more expansive mindset, which helps with creativity. What this means is that if you are trying to be creative, you might want to actively visualize what you're trying to do and add things to the picture in your head, trying them out. But it's not that subtractive imagination is bad, it's just that it's good for different things. Keith Markman, the psychologist who ran these studies, found that using subtractive imagination aided in performance on analytical problem-solving tasks, which require more straightforward, step-by-step thinking.[60]

A counterfactual mindset, by doing subtractive imagination (for example, "What if I hadn't had that fourth beer?") can also help with creativity problems that require association. Association is finding links between things. Sometimes those links are not obvious, such as the idea of using a tack box to hold a candle, but it's not coming up with something particularly new. So imagining how things would be without certain elements helps with one of the most popular tests of creativity, the Remote Associations Test, which basically tests your ability to find connections between three apparently disparate words, such as "age," "mile," and "sand" (the commonality is "stone.") This kind of thinking also helps with analytical problems, such as those on the LSAT (the Law School Admissions Test).

Problems that require thinking of something new, called "generation tasks," such as coming up with the name of a new product, are actually hindered by the counterfactual mindsets generated with subtractive imagination. Laura J. Kray found that people who were put in the negative counterfactual mindset (caused by doing subtractive imagination) came up with ideas that resembled the examples given and were rated as less creative by judges.[61]

On the other hand, the mindset generated by using additive imagination (for example, "What if I'd brought peanut butter along?") has the opposite effect! It hinders performance on association tasks and help with generation tasks. Apparently the expansive thinking style encouraged by additive counterfactual thinking helps in generation.[62] It's interesting that imagination can help creativity in different ways depending on whether you are encouraged to add or subtract elements in your imagination.

IMAGINATION AND SCIENCE

Some of the most spectacular uses of imagination are in scientific discovery. Visualization, in particular, has proven helpful for the practice of science, in part because imagination is like a physical simulation that allows scientists to rapidly "test" ideas for plausibility in their minds.[63] Scholars have found that visual imagination played important roles in many historical scientific discoveries, from DNA to Maxwell's electromagnetic field theory.[64] David Uttal did a meta-analysis of studies and found that spatial skills, in particular, predict people's achievements in science, technology, engineering, and math.[65]

One thing imagination and imagery can be good for is helping us make analogies. There's a famous experiment in psychology that that goes like this: A person is presented with a story about an army attacking a fortress. There are several roads leading to the fortress, and the general learns that the mayor has placed mines in the roads. However, because the mayor needs to use these roads for trade and travel, each mine was engineered so that it would only go off if a large weight, like that of a whole army, is on a single road. So the general cannot use his original plan, which was to bring the whole army down a single road. He splits his army up into many small groups and puts each one on a different road. Then they converge on the fortress at the same time. Thus the whole army attacks the city at once, but they don't set off any mines, because each army subgroup was too small to trigger them.

After people hear this story, they are presented with a new problem: A patient has an inoperable tumor deep in their body. There is a laser beam that can destroy it, but the problem is that the beam is so powerful that it will also destroy healthy tissue on its way through and kill the patient. How can this problem be solved?

Some people are able to use the first story to solve the problem: they suggest using several beams, each of which too weak to do any damage, but that converge at the tumor, destroying it. Much like how the general split up his army, the doctor can split up the energy in the laser.[66]

Although psychologists tend to be surprised at how few people can solve this problem (using the analogy) without a hint, to me, it's amazing that anybody can do it at all, when you think about the information processing required. Look at the differences between the two situations: to make the analogy, the person needs to be able to see the similarity between an army and a laser. These are very different things. For the mind to see that they are similar, it needs to gloss over a lot of details. What do they have in common? Well, both have the function to destroy something. They also might look visually similar in your imagination, if you imagine them both as, say, arrows pointing to something in the middle.

The problem is even more complicated when you think about the transformation that needs to be made. In the army case, you have some finite number of troops, and you can divide them up into smaller groups, depending on how many inbound roads there are. But you can't do this with the laser beam, because it can't be broken up into discrete groups the way an army can. The same amount of energy needs to be used to make a bunch of weaker lasers.

It's possible that visualization can help with this, too. If you imagine a strong arrow as a thick one, you might imagine a weak arrow as a thin one. Breaking up one thick arrow into several thin ones can apply equally to both the army and the laser, perhaps helping people make the jump of applying the transformation of one to the other.[67]

I bring up this analogy because it has similarities to the kinds of analogies that actual scientists have used to make scientific discoveries.

Although imagination has been very important for scientific discovery, it can also hold us back. Philosopher Daniel Dennett quoted one of the fathers of modern genetics, William Bateson, while speculating about the biological basis of inheritance in 1916. He was apparently unable to imagine how materials in the body could provide a code for inheritance: "The properties of living things are in some way attached to a material basis, perhaps in some special degree to nuclear chromatin [chromosomes]; and yet it is inconceivable that particles of chromatin or of any other substance, however complex, can possess those powers which must be assigned to our factors or gens [genes]. The supposition that particles of chromatin, indistinguishable from each other and indeed almost homogeneous under any known test, can by their material nature confer all the properties of life surpasses the range of even the most convinced materialism."

Reading this, we can see that Bateson was unable to *imagine* anything like DNA actually working, and made a conclusion about what the world must be like based on his confidence in his own imaginative abilities. Even writing what DNA actually is makes it sound incredible and otherworldly: billions of base pairs in every cell in the body.[68] Can we blame Bateson, given what he knew at the time, for his failure to imagine the possibility of how something like DNA might exist? But at the same time, DNA does exist. He referred to the results of his imagination and felt convinced that his inability to imagine a solution showed that something like DNA was impossible.

In fact, even now, we can't make a mental image of three billion base pairs. What I mean by that is we can imagine the *concept* of three billion somethings, but we certainly cannot visualize it or see it in our mind's eye. Picture something—it doesn't matter what—a triangle, or a dot of light, an ocelot—and now picture two. Now three. Keep adding more until you can't picture them all at the same time anymore. What number did you end up with? It might have been four, or it might have been nine, but I'm willing to bet it's a number fewer than three billion (Ryan Scoville, a particularly bright friend of mine, claims he can imagine twenty).

Because our mental images are based on memories of our perceptions, they tend to be limited by them. But we only can directly perceive a small slice of our world. This can be a problem for scientific imagination.

Our perception takes energy of some kind from the environment and transforms it into representations in our heads. Some of these representations are fleeting and vanish almost as soon as they are created. Some are further changed into other representations, and some, which we tend to call memories, are stored for later use.

So the first filter on what we perceive is our sense organs—our eyes, ears, and so on. These organs, and the brain areas that process the information from them, evolved to help us survive and reproduce in our environment. If being able to perceive something didn't help our ancestors survive and reproduce, we probably didn't evolve ways to perceive that thing. A great example is ultraviolet light. It certainly exists, but we simply can't see it. But many animals can. Lots of flowers have beautiful ultraviolet patterns on them that we're completely unaware of. These patterns are signals to insects, many of which *can* see ultraviolet light. The electromagnetic spectrum is wide, and the range of values in it that we can see with our eyes is really small. X-rays, infrared, gamma rays, and ultraviolet are all colors, in some sense, just as much as red is. But because our ancestors had no real need to see these frequencies on the spectrum, we didn't evolve to be able to see them. And as a result, we don't imagine images painted in gorgeous shades of ultraviolet.

The situation is similar with sound. Dogs, for example, can hear higher pitches than humans can. These are relatively simple kinds of energy (electromagnetism, kinetic waves in the air) that we can't naturally detect. But the problem goes deeper.

We evolved to deal with situations in what biologist Richard Dawkins calls the "Middle World." This means that we only really have to deal with medium-sized objects. In physics, the set of "medium sized objects" range from things bigger than drops of water to planets. Granted, this is a pretty wide range of things.

But there are lots of "objects" in our universe that are not medium-sized. Our common sense, and our perception, is exquisitely tuned to make sense of Middle World, which is why our common sense breaks down when faced with the science of the very small (such as quantum effects), very large (the behavior of galaxies), very slow (forest growth, tectonic plate movement), or very fast (relativistic effects as something approaches light speed). For example, in quantum mechanics, it's possible for a particle to be in two places at once. This makes absolutely no sense in Middle World, so we tend to have a knee-jerk skepticism to the very idea that it's possible at any scale. Quantum mechanics appears to be full of paradoxes, because it violates common sense—but common sense is a product of what we were evolved to be able to see.[69]

For example, solid objects are made of atoms, and atoms are, for the most part, "made of" empty space. If an atom were the size of a sports stadium, the nucleus of an atom, where most of the mass is, would be about the size of a house fly in that stadium. But our experience of an object, such as a rock, is of solidity. Why? Because our hands can't pass through it. If we were the size of subatomic particles, Dawkins muses, we would perceive rocks quite differently: what appears to constitute "reality," for a given species, is what it *needs* to be. We perceive rocks as solid because when we interact with them, they behave as though they were. When we step in a puddle, our foot goes right through. This is not so for some small insects, who stay on top of the water because of surface tension. Humans don't really think of surface tension because it's mostly irrelevant to how we interact with liquids. We're too big for surface tension to matter.

Because we understand the world in this way, we store information about our world accordingly, and our imaginations reflect it. Try as we might, it's very difficult to imagine, say, the ten spatial dimensions of the branch of physics called "string theory." We can imagine three dimensions just fine, but have trouble even with four, let alone ten. When we imagine, we tend to imagine Middle World things. It's easier.

Let's dig into this a little bit further. Can you imagine a square with five sides? Well, you kind of can, and you kind of can't. In one sense, it's easy to imagine the idea (but not a picture of) a square, and just attribute to it five sides instead of four. It might be inconsistent with your other beliefs, but we have no trouble imagining contradictory situations. However, actually forming a mental *picture* of a five-sided square is impossible. Or, if you think you can do it, I'd love to hear about it.

Even our sensory imagination is limited to the ranges we're used to experiencing. Imagine a musical note—I don't care which one. Then start going up the scale: CDEFGABC and so on. How high can you go before your imagination becomes muddled and you can no longer distinguish or "hear" the note in your mind's ear or distinguish it from the note just below it in your auditory imagery? I'm not saying that you can't go pretty high—perhaps even higher than any note you've ever heard—but I am saying that there's a limit to what you can imagine. (A casual canvass of my musical friends suggest that they can only imagine three to four octaves.) You can try the same exercise with redder and redder reds, or longer and longer arms, or a beep that lasts for less and less time, or a flower that gets more and more beautiful, or a taste that gets more and more peanut-buttery. At some point you reach your limits of your own imagery.

What this all means is that although our imaginations are vitally important for scientific creativity, mental *imagery* might actually hold us back when trying to make theories about things that aren't a part of Middle World. Just like a congenitally blind person cannot visualize, there are things we can only understand conceptually, and often with difficulty.

We can only imagine non-Euclidean geometry conceptually, not with imagery. Galileo's finding that gravity makes objects fall at the same speed, no matter their size or mass, is very counterintuitive because in the world we live in day-to-day, air friction makes really light things fall more slowly.

Whether or not there are things in the universe so strange, so unlike Middle World, that we cannot even *conceptually* conceive of them is an open question. If there are, then we might have to wait until we have some superior intelligence (perhaps human intelligence augmented through technology or genetic engineering, or have artificial systems that don't have our same constraints), before they can be understood.

9

Imaginary Companions

Did you ever have an imaginary friend?

Imaginary companions, commonly called "imaginary friends," are common for children. Sixty-five percent of children under seven play or interact with one on a regular basis.[1] Some kids love their imaginary friends even more than their real ones.[2] The classic imaginary companion is completely imaginary, but many children play with imaginary versions of their real friends, of characters from movies, or imbue stuffed animals with humanlike traits. A scrappy old toy might be a beautiful large animal in the mind of the child.[3] One child was interested in dolphins, so her parents gave her a toy dolphin. "Dipper," in the mind of the child, was large, and sparkly, and from a star.[4] One child had long conversations with a chest of drawers.[5] As in regular pretend play, such as imagining a banana as a phone, the child has no problem distinguishing an object in the real world and its fantasy counterpart.

Where do we draw the line between true imaginary companions and the temporary animacy children give to toys in a brief act of pretend play? Many children carry around stuffed bears, but unless the child imbues the creature with humanlike traits and treats it as a real friend, scientists won't count it as an imaginary companion. Kids might be asked this question: "Pretend friends are like dolls or toys that you pretend are real or people that you pretend are real. They are make-believe friends. Do you have a pretend friend?" In response to this question, about 50 percent of children answer yes. Of these imaginary companions, 39 percent were based on toys or stuffed animals, 19 percent were invisible animals, and 41 percent were invisible people.[6] It's easy to underestimate the frequency of this phenomenon because parents don't always know that their child has one. Marjorie Taylor, a major researcher in this area, found that after age four, only seven of the thirty-two imaginary companions found in a study were known of by the parents.[7] On the other hand, the project is complicated by the fact that children will sometimes spontaneously invent an imaginary friend when asked about one, or invent details on the fly as they are asked about the ones they have![8]

Taylor tells of the personalities attributed to the childhood stuffed animals of one of her graduate students. The graduate student was aware of the possible social ramifications of having so many stuffed animals: she would take care that they were good to each other, so they would not retaliate with envy when another one got to go on a trip with her. She would carefully introduce new animals and reassure all of the others that they were still loved. As a child, this graduate student agonized over which stuffed animal got to go on a trip with her—her parents only allowed her to bring one.[9] We can see from this example that children practice what they are learning about social relationships in their pretend play.

Some children imagine *themselves* to be the imaginary companion. Of course, during pretend play children often take roles. We only consider role-playing to be in the category of imaginary companions in those cases where the role is more complex and persistent over time—for example, if a child pretends that she is Wonder Woman every day for a month.[10]

This kind of role-playing can be annoying for parents. For example, sometimes kids will decide that they are cats, only eat out of bowls on the floor, walk on all fours, and look quizzically when asked to stop—the company be damned.[11]

So how long do imaginary companions last? If they are too brief we don't call them imaginary companions at all. Some can last for years. Some children have many who come and go from their lives, and some companions are passed down from sibling to sibling in the same family![12]

Many imaginary companions that appear to go away actually just become secret. Many parents in the English-speaking world believe that it's not normal to have imaginary friends after about age five or seven—in fact there's no evidence that they are harmful at any age, so this is just a myth. When the child perceives that the parent starts to disapprove, the imaginary companions go dark: the children stop sharing information about their companions and only play with them when the parents aren't around. During middle childhood, parents start knowing less and less about the fantasy lives of their children. If the people around the child don't tolerate the companion, it might go "underground" or disappear altogether.[13]

But as important to children as their imaginary companions are, when they disappear it is done without ceremony or fuss.[14] Children might simply state that the friend "died," or just not mention the friend again. Often old imaginary companions are replaced with new ones.[15]

Many children simply can't remember the imaginary companions they've had and dismissed and often show little interest in talking about them—particularly if they get a new one.

About half of adults surveyed say that if they had companions, they disappeared by about ten years of age.[16] But surveys of adolescent diaries found that older children have them, too—28 percent of the sixteen- to seventeen-year-olds. When adults have imaginary companions, they are sometimes seen as a sign of mental illness. But as far as I can tell, there is no evidence that imaginary companions in adults is harmful, or even a symptom of something else being wrong.

That said, we should take care to distinguish imaginary companions from hallucinations. The important difference is that unlike many hallucinations, people don't actually believe in their imaginary companions—they have insight. Atheists will sometimes refer to God as people's imaginary friend, but the belief in God or gods seems to be very different. People who believe in ghosts, angels, demons, spirits, or gods really think they exist, even if they cannot be sensed directly. For these reasons, beliefs in gods are considered a phenomenon different from that of having imaginary companions.

As children age, their fantasies can become more complex. Rather than simply having imaginary friends, they might create entire worlds, called "paracosms," complete with languages, cultures, and so on. A study by Robert Silvey and Stephen MacKeith found that most paracosms were created around age nine, and were usually abandoned by age eighteen, but this was an unsystematic study of friends and acquaintances. It was conducted between 1932 and 1968,[17] long before the invention of tabletop role-playing games such as *Dungeons & Dragons*. I know from personal experience that playing these games encourages children to create large, complex imaginary worlds. Before I graduated from high school, I'd created several paracosms, and have even more since then. Speculative fiction writers also create complex worlds, but because these authors are relatively rare, it makes sense that they would not appear in a small, representative sample of the population. Nevertheless, Taylor ran a more recent study of twenty-two adults and found that five of them reported having imaginary worlds complex enough to count as paracosms.[18]

AUTONOMOUS CHARACTERS

Many fiction writers say that their characters seem to have minds of their own. Writers sometimes report that they feel that the events in their novel, or even the words themselves, are being dictated to them outside of their conscious control.[19] Some writers report that they need their characters to do something, presumably for some plot reason, but the character "refuses" to do it. This feeling, the "illusion of independent agency," is quite common. I was at a writers' panel and one author said that her characters wouldn't do what she said, and another writer said that he was in complete control of his characters. Marjorie Taylor surveyed fifty fiction authors and found that a full 92 percent of them experienced this phenomenon of their characters having their own agency.[20] Some writers even report that writing feels more like dictating what their characters do and say than creating the story deliberately.

Some characters feel so real authors have imaginary conversations with them, much like children have conversations with imaginary friends.[21] John Foxwell's research found that 69 percent of authors hear voices of their characters, and 42 percent can enter into dialogue with them. Sixty-five percent say they can act on their own accord.

Mary Watkins has documented evidence of many famous authors who insist that their characters are autonomous and out of their control. For example, Philip Pullman, author of the His Dark Materials trilogy, reports negotiating a deal with his character Mrs. Coulter to get her to spend time in a cave in one of his books. Some authors have reported that their characters give them unsolicited advice about the writer's own life![22]

Seventy-one percent of authors say their characters become more autonomous over time. After about thirty thousand words, some of the main characters seem to take on lives of their own. One author said, "I nowadays just plan my books halfway as I know that in the middle of the writing process the characters will take over the story so my planning will be useless anyway."[23]

The perceived autonomy of these characters is very strong. Some authors even insist that one can't be any good as a writer unless this is the case. One author said that without their main characters becoming autonomous, the characters become shallow archetypes.

But sometimes it can get in the way of the story the author is trying to write, when their characters refuse to behave and do what is needed for the book. One author said: "They develop their own narratives, rapidly accumulating their own histories and anecdotes—if unchecked, I have had to kill off characters to stop a story digressing out of sight."[24] When Alice Walker was writing *The Color Purple*, not only did her characters seem to choose their own actions in the plot, but they regularly visited her and commented, sometimes unwelcomely, on Walker's own life.

When children feel that their imaginary friends have minds of their own, it could be like fiction authors who experience the illusion of independent agency—their imaginary friends are so familiar to them that they no longer feel like their own creations. Intriguingly, the writers who experienced the illusion were about twice as likely to have had imaginary friends as children.[25]

We don't even need to turn to imaginary companions or to fiction authors to find imagined characters acting with their own agency—most of us experience such characters every time we dream: we have imagined experiences of vivid characters we can't consciously control. This is even true for lucid dreamers. Composer Robert Schumann believed that his compositions were communicated to him by other (dead) composers, such as Beethoven.[26]

What is going on here?

Just as authors sometimes complain that their characters will sometimes act in ways they do not like, imaginary companions can be very annoying to the children that create them. At the core is the issue of control. Even though imaginary companions are make-believe and the people who have them know it, sometimes the friends feel out of control, or can even be downright frightening. One fourteen-year-old boy slept with his hands covered for three years out of fear that a giant would step out of a picture on the wall to cut off his hands.[27] It's unclear whether this giant counts as an imaginary companion, because he does little more than scare the boy, but the normally friendly imaginary companions of children can not bother to accompany a child on a trip, sometimes fail to show for playtime, talk too loudly, not share, and so on. One child's day was ruined when his imaginary companion, a pony, had made other plans and could not accompany the child to a horse show.[28] Jennifer Mauro found that 34 percent of children with imaginary companions say they get angry with the children sometimes, and parents witness arguments between the child and the imaginary friend.[29]

But most of the time children are in conscious control. Marjorie Taylor tried to get children in the lab to be envious of their imaginary companions. They gave the companions a present and informed the children that the companion refused to share. The experiment had to be discontinued, because the children got annoyed with the researcher rather than with the companion! They don't abide other people trying to control *their* companions.[30]

So why do children sometimes seem to be in control of their companions and sometimes not? Why are some characters feel in control by their authors and other times not? We distinguish hallucination from other imagination, in part, because we are in control of what we imagine—in hallucination and dreaming, for example, we have little or no control. So if these children are simply imagining their friend, why not simply make the friend behave? Somehow, the friend seems, to the child, to have its own autonomy.

It could be that, in the cases of imaginary companions and well-fleshed out characters that authors imagine, the person's idea of what the character is like is so detailed and well-understood that the mental processing done to explain and predict what these characters say and do becomes completely unconscious. It's not that the character is out of control of the person who imagined them, but they are out of control of the *conscious part of the mind* that created them. The character's actions are determined by the deep tides of the unconscious ocean of their creator's mind.

To understand my interpretation of this phenomenon, it's important to understand a pervasive ability of the human mind known as "automatization," or making things automatic. When you do something complex for the first time, it's difficult or downright impossible—recall learning to drive, playing pinball, or dancing salsa. What happens is that over time the task becomes automatic. You can then do the task pretty well, often without even thinking about it.

You might have experienced meaning to stop at a store on the way home from work, but then finding yourself home, without having thought about the ride at all during your trip. While you were driving, you were thinking about other things. You might have started out with the intention to stop at the store on the way home but ended up not doing it. Your automatized memories got you home.

When I was first learning to swing dance, it seemed impossible. If I paid attention to my feet, my hands did the wrong thing. If I turned my partner correctly, I'd lose the beat and not know where to put my feet. But after doing it for many years, I got to the point where I could do the basics without any conscious control at all. I could think about what was coming up in the song, so that I could time a dip perfectly. I could hold a philosophical conversation while swing dancing at a fast clip. And I met my wife this way. Such is the power of automatization. It can help you get married.

Activities get so well learned that they become fast and unconscious. It is most obvious in physical tasks, such as sports and driving. It could be that automatization can also happen to reasoning—that is, doing things that don't just involve moving your body. Sometimes, for example, I hear an argument from someone, and I instantly know that there's something wrong with it. When asked to elaborate, I figure out why the argument was wrong as I talk it out, and usually my feeling that there was something wrong with the argument was right on. But what's interesting is that I was convinced that it was wrong before I was conscious of the reasons why! What is probably happening here is that there is an automatized, unconscious reasoning going on, and only when I have to talk about it am I required to use my conscious mind to reconstruct, or invent, a justification for the feeling.[31]

Fiction authors need to create whole people in their minds, including appearance, mannerisms, culture, ways of speaking, attitudes, preferences, personalities, goals, and so on. At first, this can be cognitively demanding, but over time, thinking about this (fictional) person becomes automatized. Because the author is no longer conscious of the thinking process that determines what the character would want, do, or say, it feels like the character is a person telling her (in her mind) what's what. It is similar to a mental models you create of people you know very well. You instantly know that your spouse would like this book or not like this restaurant. It could be that the same trend is happening with imaginary companions and novelists' autonomous characters[32]: they start out as a conscious creation, but then when they become automatized, they can start "misbehaving."[33]

So what might be happening with these authors is that they start out consciously thinking about a character, and after they get a good mental model of what the character is like, it becomes automatized, so that they no longer need to consciously reason out what the character would do in a particular situation, how they would react to things. Just like all automatized things, it feels like it's coming from nowhere. The characters are being thought about by the old brain, unconsciously.

The new and old brain distinction is important for understanding imagination and creativity, because often we have no idea where our creative ideas come from. The old brain is largely unconscious, so ideas that bubble up from its depths seem to come from nowhere. Long before a complex understanding of the unconscious mind, the ancient Greeks attributed these ideas, which seemingly came from nowhere, as being delivered by gods, the muses.

But not all of these unbidden ideas come from ancient evolved systems. Because we can automatize practiced actions, we can render them unconscious, too. My ability to swing dance without thinking about it certainly isn't an evolved behavior!

So unconscious processing is a combination of evolved abilities, and formerly conscious abilities that have been made fast and automatic. Keith Stanovich calls this unconscious processing the "Autonomous Set of Systems." When we get an unbidden idea, we know it's from the Autonomous Set of Systems, but we don't know whether it was a well-learned or evolved response. In this book, when I refer to the old brain, I really mean the Autonomous Set of Systems, because I want to include not only evolved processes, but well-learned ones as well.[34]

TULPAS

We can see how some kinds of imaginary companions can be generated through learning and effort in the community of people who create *tulpas*. The idea is that a person, called a "host" or a "tulpamancer," will use mental exercises to eventually generate a being with whom the host can interact in their imagination. It actually got popular in the West by adult fans of the *My Little Pony* franchise (known as "bronies"), who wanted to create pony tulpas to interact with. They can be thought of as deliberately created, semipermanent, nonpathological hallucinations. They are thought to have their own sentience, separate from the host's.[35] Some tulpamancers actually believe that their tulpas are supernatural in some way, but many believe it's just a psychological construct.[36]

My (speculative) automatization theory, then, suggests that with lots of practice, the mental model for the imagined person becomes subconscious, which explains why it feels like they're acting autonomously. This explains the attribution of autonomy to tulpas, which reportedly requires about four months of practice in "the forcing stage" to generate one. Occasionally, tulpamancers will try to undo a tulpa, and this is supposed to be very hard, like trying to unlearn how to play a musical instrument.[37] This suggests, to me, automaticity by learning. Similarly, religious revelations can take hard work to train yourself to hear voices attributed to God.[38] We have also seen that fiction authors' characters do not start out autonomous, but rather the autonomy appears only after long and careful thought of what the character is like, and only tends to happen to main characters.

But the theory doesn't shed much light on dream characters and hallucinated characters, who seem to be autonomous from the get-go, requiring no practice or learning period at all to get the perception that they have minds of their own. My explanation for this is that the imagined characters who seem automatic from the beginning are stereotyped, and the slow growth of autonomy of other imagined characters, who have more fleshed-out personalities, requires practice.

What about imaginary companions? We actually don't know whether they start out as characters under conscious control and change over time. By the time researchers talk to kids about their companions, they have had them for many months[39] and, believe it or not, five-year-olds are not great at questions like, "Do you experience your friend as an autonomous character?"

Dream characters seem to be autonomous (even lucid dreamers tend to report autonomous dream characters). We also know that your prefrontal cortex, the new brain that does deliberative processing, is relatively quiet during dreaming. So your mind is probably unable to deliberately choose how some character is going to behave, like a fiction author might in the early stages of writing a novel. Suppose you dream that somebody is threatening you and trying to hurt you. This need not be a deep character, in terms of motivations, preferences, moral code, or anything else. They serve a function in the dream, and that function is to threaten you, in the context of the other dream elements. Similarly, if you walk down the street in a dream, the people you pass also feel autonomous, but not complex. I suggest that dream characters tend to be stock, stereotyped characters, without much depth.

Now dream characters of real people you actually know (like your romantic partner, coworker, or family member) can be very complex. Doesn't this violate the theory? No, because you also have a detailed mental model of these people from interacting with them for so long in real, waking life. Your ability to predict how they might behave, then, could be automatized.

What about hallucinated characters? Those with Charles Bonnet syndrome are certainly very simple. They don't even talk. Most schizophrenic hallucinations are voices in the head, but do not seem to be lasting characters of any complexity. As for multisensory hallucinations of people, I don't know of any research in it, but I would expect that such characters would either be simple or based on real people the hallucinator knows.

Let's get back to imaginary companions: so imaginary companions can be controlled by the old brain, and sometimes they can do things that make the child upset. But is having an imaginary companion a problem? Or a symptom of another problem?

Lots of people think that there's something wrong with children who have imaginary companions and that, perhaps, they are a sign that children can't tell the difference between fantasy and reality. Even parents who think imaginary companions are okay when the child is really little think that there is an age where it's not okay anymore. A long time ago, there were studies suggesting a negative picture of imaginary companions, but they suffered from methodological problems.[40]

So is it a problem? First, let's look at the kinds of children who are likely to create imaginary companions. They are just as likely to be boys as girls, although boys tend to create them later and are more likely to embody them—that is, they are more likely to impersonate the character, rather than seeing them as a separate entity that they play with.

Various studies on the subject show that these children are more cooperative with adults, engage in more cooperative play, are rated to be less shy by their parents (though other studies have failed to replicate this), are less fearful and anxious about playing with other children, are described as smiling and laughing more than children without imaginary companions, and have better focus (up to age seven, after which the differences disappear).

They are also better at "theory of mind" tasks. This means that they are particularly good at imagining what might be in the minds of others. For example, suppose you show a child, Pat, that what looks just like a rock is actually a painted sponge. Then you ask Pat whether a new person coming into the room will think the object is a rock or a sponge. If Pat has a good theory of mind, she will say that the new person will think it's a rock—even though Pat, herself, knows better. However, if Pat's theory of mind abilities are not yet developed, or are somehow compromised, she will be unable to imagine that anybody could think differently than she. In this case, Pat would say that the new person would know it's a sponge.

In terms of personality, some studies have found differences, but the overall finding when comparing the personalities of children with and without imaginary companions is that the similarities are more striking than the differences. There also appear to be no differences in intelligence or creativity. However, children who are prone to fantasy (which includes, but is not limited to, having imaginary companions) appear to be more creative than other children, according to certain creativity measures: they tell more novel stories given a set of verbal stimuli and are freer in their imaginative play.[41]

It seems that pretend play of all kinds is hardwired into human nature. Adults who report that they had imaginary companions are less neurotic, less introverted, more dominant in face-to-face interactions, more self-confident, and more sociable.[42]

FANTASY VERSUS REALITY

Some worry that children with imaginary companions have trouble distinguishing fantasy from reality. For example, sometimes the child and the imaginary companion will partake in a game that gets scary, and the child will retreat from the game in fear. Isn't this evidence that the child thinks it's real? If the child really knew it was just imagination, then the child wouldn't be so scared, would she?

Sure she would.

Adults have experiences very similar to this when watching scary movies. Once the movie crosses the line from exciting to very unpleasant, the adult will stop watching the film. Or how about when you ask someone to think about something terribly sad, like the thought that they might stop making Star Wars movies, and they say they don't want to think about it because it would be too upsetting. Don't they *know* it's not real? Do the emotions adults feel mean that they can't distinguish fantasy from reality either?[43]

The answer is kind of yes and kind of no, for both children and adults. The new brain knows full well that imaginary companions and horror movies are depictions of things with no physical reality. But the old brain can't tell the difference. It responds to images of people in movies just like images of real-life people. To an extent, it also responds similarly to our own mental simulations. So in both cases—children and adults—the emotional response people have is not a reliable indicator of people's conscious ability to distinguish fantasy from reality. Part of your brain thinks it's real, and part of it knows it's not.

Overall, children's abilities to understand the nature of "pretend" and fantasy is quite impressive. When it comes to imaginary companions, sometimes, after a scientist asks the child many questions about the imaginary friend, writing everything down, the child will worry that the adult is taking it a bit too seriously and say something like, "You know it's just pretend, right?"[44]

What is particularly amazing about it is how they are able to do this when, in our modern world, kids are absolutely inundated with fantasy! From books, to television, to Santa Claus, to their own pretend play, a great deal of what kids pay attention to isn't real at all. Adults even play along, talking to stuffed animals as though they were real or treating a pile of cushions as a house.[45]

And unlike stories given to them in movies, imaginary companions, like other forms of pretend play, are completely controlled by the children who create them, albeit sometimes unconsciously.[46]

By age five, kids seem to have little trouble distinguishing fantasy from reality.[47] Three-year-old children are 75 percent correct when asked to sort things into boxes labeled "real life" and "make-believe."[48] Interviews with parents and children show that nearly all children understand that their invisible friends are pretend.[49] Studies show that children who have imaginary companions, or play fantasy games, or have lots of dreams, are no more likely than other children to believe in fantasy characters.

According to Marjorie Taylor, kids' confusion about fantasy and reality are most prevalent when adults are actively trying to deceive them![50] It is ironic that many of the parents who worry about their children's ability to distinguish fantasy from reality are the very same people who go to extraordinary lengths to convince children of the reality of culturally accepted fantasy characters, such as the Tooth Fairy, the Easter Bunny, and Santa

Claus.[51] My own parents would help me put out cookies and milk for Santa, then eat some of the cookies while I slept, and then point to them in the morning, citing them as evidence of Santa's existence. As a young child I found this utterly convincing. I wasn't alone. Among families who celebrate Christmas, 85 percent of four-year-olds strongly believe in the existence of Santa, and the percentage only goes down to 20 percent by age eight.[52] It seems quite unfair to blame children for their inability to distinguish reality on the basis of cultural fantasies we try so hard to convince them of! We should be more impressed with their ability to figure out the truth in spite of all the fantasy we flood their heads with.[53]

I should also add that sometimes healthy adults confuse fantasy and reality, too—as when actors who play doctors on television get letters from people asking for medical advice.

Sometimes parents are concerned about imaginary friends and other characters on religious grounds. Many believe that culturally accepted imaginary characters (such as Santa Claus and the Tooth Fairy) will ultimately undermine religious belief when the children find out that they are not actually real. Indeed, children tend to think of Santa and God as connected in some way. But a study by Cindy Dell Clark failed to find evidence that children lose their faith in God when they learn the truth about Santa. But parents still worry about it.[54]

Although some parents tell their kids there is no Santa, kids tend to believe in him anyway. Clark surveyed a bunch of kids and found that all but one of the children of fundamentalist Christian parents believed in Santa, in spite of what they learned at church and from their parents.[55]

In some cultures, imaginary companions are feared because they are thought to not be imaginary at all. Many fundamentalist Christians are wary of fantasy because they believe that Satan can influence children in the form of imaginary companions. In a sample of Indian families, none reported having children with imaginary companions. However, when children talk to entities that are not there, they are often referred to as "invisible," and are thought to be real entities in the spiritual realm or one of the past lives of the children—and, thus, not imaginary at all. However, the rate of invisible entities in India is much lower (0.2 percent) than it is in the West (about 30 percent). Why there is a cultural discrepancy is unknown, but one thought is that Indian children have less time alone than Western children typically do, lessening the need for imaginary company.[56] As we will see, many kids have imaginary companions for company and to have someone to play with.

SO WHY HAVE AN IMAGINARY COMPANION?

If they're not doing any harm, are they doing any good? What function are these imaginary friends serving?

Marjorie Taylor lists many functions of imaginary companions. They can help the child express anger without hurting a real person's feelings. The companion can be used as a scapegoat and blamed for naughty things the child did. Sometimes children will argue that they should not have to do something because their imaginary companion does not have to. Children can express an opinion or fear through the imaginary companion ("Noogy is afraid she'll freeze if she plays outside."). Sometimes kids will soothe the fears of their companions as a way to soothe themselves. Often the companion disappears when the fear disappears.[57]

Ultimately, interaction with imaginary companions is a form of play, and like other forms of play, it seems to help children control their emotions.[58] Imaginary companions can be used to help deal with traumatic events, such as war or medical issues. Rather than being viewed as a problem, imaginary companions are often a form of self-therapy. But most imaginary companions are created for reasons other than abuse.[59]

You might think that having an imaginary companion has something to do with what is popularly known as having a "split personality." In psychology, this disorder is called "dissociative identity disorder," and involves more than one personality in a single person. Let me first say that just about everything we think we know about dissociative identity disorder, including its very existence, is debated in psychology. That said, in general, it seems to be unrelated to imaginative play in children, and there are some striking differences between a dissociative identity (or alternate personality) and an imaginary companion. Companions do not often take over the body of the child, and when they do, children with companions are completely aware and in conscious control of what's happening. However, sometimes a dissociative identity can be traced to what was once an imaginary companion created to help deal with an abusive situation. But only a subset of people with dissociative identity disorder had imaginary companions as children.[60]

Although children create imaginary companions to help deal with their problems, clinicians observe that children who are victims of abuse or neglect, in general, have *inhibited* creativity and imaginative play behaviors. Their play instead is characterized by monotonous tasks, done with little joy.[61]

Children also use imaginary companions to aid with building skills. Some children have companions who are completely unable to care for themselves. The child must "take care" of the companion. The child herself feels competent in contrast. Other children create companions of great strength and intelligence, perhaps offering inspiration or a powerful ally. Girls are more likely to create the incompetent companions, mirroring the kinds of toys they tend to play with (baby dolls versus action figures). Recall that boys are more likely to act as their companions, too. It's probably more fun to act the role of a competent rather than an incompetent character, suggesting that these gender tendencies (enacting the character or not and having a superhuman versus a dependent companion) might be related.[62]

Children who have imaginary companions watch less television, which is not surprising because many kids report watching TV for the same reasons they report playing with imaginary companions: it's fun and provides company. Companionship and having fun are the primary function of these imaginary friends.[63] This explains why children often create them when reality does not provide them with enough playmates. Often (but not always) the imaginary companion is only around when the child is alone.

CULTURAL DIFFERENCES

Most parents view imaginary companions as harmless, at least until a certain age. But if and when a parent sees them as strange or unhealthy, children simply hide the companions from their parents.[64] JoAnn Farver found that Mexican mothers, unlike American ones, do not seem to see pretend and imaginative play as important to the education and development of their children.[65] And Maria Montessori, who created the theory behind Montessori schools, had a focus on intellectual exercises and excluded *all imaginative play*. Traditionally, Montessori classrooms have no toy animals, dolls, dress-up costumes, or any of the other playthings that children like to use when doing pretend play.[66]

Between 1981 and 1997, the time children spent on free play dropped by 25 percent.[67] This is a shame, because play is vitally important for development, whether you're a little kid or a puppy. It is critical for learning how to cope with stress, to learn social and cognitive skills. Kids learn to take turns, because not doing so will probably lose them a play partner. It can even make kids smarter. Free play differs from already-structured games, such as sports, and involves more creativity.[68] It also correlates with better imagination, creativity, and even math ability![69]

Education in the Western world has grown more and more structured, with specific lessons and tests. There is less and less free-form, pretend play, and creativity might be suffering as a result of this.

To summarize, the existence of an imaginary companion does not appear to be a problem in itself. However, in certain cases it might be a sign of another problem, but in these cases the companion is helping, not hurting. Children are more likely to generate them and to engage in pretend play in general, if their lives are not overly scheduled. Children need time, props, space, and sometimes even boredom to encourage the development of imagination, and imaginary companions are a helpful manifestation of it.[70]

10
Imagination and Technology

Historically, we've only been able to know about the content of people's imaginations to the extent that they were able to tell us about them. As a form of scientific measurement, this has a lot of problems. People forget what they imagined, particularly for dreams. They might not be able to describe in words what they imagined. They might be embarrassed to describe the content of their fantasies. But what if we could scan their brains and read their imaginings right from the brain scan? Can we read someone's mind with a machine?

The short answer: not yet. We are at the very beginning of reading people's imagery with machines. Until recently, the idea of being able to read your mental images from brain activity was strictly science fiction. But in recent years science has made progress on exactly that. Using functional magnetic resonance imaging, a brain-scanning method also known as "fMRI," scientists have been able to "decode" activity in the visual cortex to know what a person was looking at in terms of line orientation, position, and even what the object was. Mark Stokes can tell which of two patterns someone is imagining based on brain activity alone, but we're far from putting someone in a scanner and generating a movie based on the pictures in their head![1] Kendrick Kay was able to tell which photograph a person was looking at (from a selection of 120) simply with the information from a brain scan, with an accuracy of 92 percent. Alan Cowen used fMRI recordings of higher-level cortical areas—that is, not the occipital areas normally associated with vision—and was able to get a computer program to learn to reconstruct faces that a person was looking at. That is, they built software that didn't just identify the face, but could create an actual image of a face a person was looking at merely by examining the use of oxygen in the higher-level cortical areas of the brain (the idea is that areas that use more oxygen are more active).[2]

That's reading the brain during visual perception. What about imagery? Neuroscientist Martin Chadwick showed people three film clips of women doing everyday activities and then had them recall one of them. Chadwick's team was able to decode which memory the person was recalling with a brain scan! All of this is particularly amazing, because technically all an fMRI can read is the oxygen use in the brain's blood vessels.[3] More recent work uses artificial intelligence to decode brain readings and map them to images.[4]

These technologies should get better in the future. But the task will be harder the more idiosyncratic each person is. That is, if your brain imagines a jar of peanut butter in its own way, different from the way everybody else's brain does it, then a single technology for brain reading won't work. Each technology would need to be trained for each individual, casting doubt on the feasibility of some nightmarish public brain-scanning camera that reads everyone's thoughts as they walk by.

We have reason to think that for the lower-level visual properties, like lines and colors, there is more consistency across people, in terms of how things are represented in the brain. But for higher-level, more conceptual imagination, people probably differ a lot. This would mean that using a brain scan to read someone's imagery (like imagining a triangle) might be easier than reading their conceptual imagination (imagining feeling sanguine about getting surgery).

But there's another way imagination interacts with technology: getting machines to have their own imaginations. This is an idea people have been talking about at least since 1962, and there's been some interesting progress.[5]

THE SCIENCE OF IMAGINATION

My own research group is called the "Science of Imagination Laboratory" (SOIL). Our goal is to understand how people generate visual scenes in their heads and to understand it well enough to replicate it with computer software. So, basically, we try to make computer programs that imagine the same way people do.

When somebody tells you a story and you imagine the scene to help yourself understand it, or you encounter a scene in a dream, or when you picture the scene of what you're reading in a novel, you're doing imagination of scenes. That's what we're trying to understand.

So what's a visual scene? The human mind conceives of objects, which are concrete, typically movable things that have an identifiable shape. Objects are made of parts (for example, a jar of peanut butter has a vessel, a lid, and the peanut butter). Objects are the components of scenes. When you ask people to list attributes of scenes, such as the beach, 95 percent of the things they list are parts or components—things like sand, sunbathers, and the ocean.[6]

So a big part of imagining a scene is choosing which objects go in them, and where they go. Interestingly, where objects go in a scene is important (for example, it's important that chairs surround a dining room table), but it's even more important for the parts of objects. If you move around the furniture in a living room, it's still a living room, but if you rearrange the parts of a deer or a car, it changes the very essence of what it is pretty drastically![7]

Suppose I told you that I walked my dog this morning, and you come up with a picture of that in your head. Some people will have a very vivid image, full of detail. Others will maybe just picture me and some vague dog, and some will get no picture at all—perhaps just the idea of me walking a dog will be strong in their mind, without any sensory-like qualities.

Let's take the case of the person with a detailed mental picture, as it will allow us to talk about more of what happens in imagination. Suppose the person pictures me in the park, in the summertime, walking a medium-sized brown dog on a leash. There is a tree nearby, grass, and maybe a park bench. This is a reasonable image of a man walking a dog. Where did our imaginer get the tree, the leash, the kind of dog, and so on? When I say "I walked my dog this morning," I did not mention any of these things. Imaginers decide, unconsciously, what objects to put in the image and what objects to leave out. There is a tree, a leash, and a bench, but no planet Saturn, or diagram of an E. coli cell, or an album that Beastie Boys would have made if Adam Yauch were still alive.

These objects must come from memory, because all imagination consists of recombinations of things in memory. Our imaginer knows that there's going to be a man and a dog already, but the other things in the image will be those objects that have been associated with the man and dogs in her perceptual history. That is, she is likely to imagine in the scene things that typically go in such scenes. So if 90 percent of dogs seen are seen with leashes, it's likely that when she imagines a scene with a dog in it, there will also be a leash in the scene.

We wanted to model this with software. We got a database of about fifty thousand pictures. The great thing about this database was that the pictures were labeled. We knew what objects were in the pictures. Specifically, for each picture, we have a list of labels that are mostly object names of things in the pictures. This database came from a game designed by Luis von Ahn, the creator of CAPTCHA and Duolingo. The game was called the "ESP Game," and here is how it was played: Players would go to the website to play the game, where they would get randomly paired up. The players didn't know who their partner was—this was crucial. Then they would be shown a picture from the web. Suppose it's a picture of a kid fishing on a boat. Each player's goal was to type in the same words as the other player. The problem is, the players can't communicate. The only thing they have in common is the picture, so they end up writing words that are associated with that picture: kid, boat, fish, sky, and so on. When both partners typed in the same word, they got points and moved on to the next picture. After three minutes, the game was over, and the players got a final score.[8]

The genius of this game is that it was fun to play, but at the same time it collected valuable information. That is, when lots of people say that a particular label like "fish" is associated with the same image, that image very likely has a fish in it. So now we have an ESP Game database of fifty thousand images with labels that are reliably associated with them. On average, there are about twelve labels per image.

The great thing about this style of game is that it has a built-in error correcting mechanism. If players make misspellings, or purposely give labels that are not in the picture, just to mess with researchers (such disreputable creatures actually exist), those labels won't be associated with the image.

So using the ESP database, we can tell what labels correlate with what other labels. Not surprisingly, "car" is associated with "road" and just about everything is associated with "man" (it's interesting to speculate on how these random images from the web show the male-dominated nature of our society).

If we think of a person's working memory as having about four or five things in it, then perhaps the top four correlated labels should be put in the image with the original label that inspired the imagining. So "dog" might pull from memory "leash," "tree," "man," and so on.

Note that this should be, to some extent, culturally dependent. In my city of Ottawa, most pet cats are kept indoors all the time. There are very few stray cats because they can't survive outside for long in the winter. So you don't see cats outside much. Dogs, however, you see outside all the time, because they are popular pets that need to be walked outside. The only time we see dogs indoors is when they belong to us or our friends. So if a person's visual history is mostly from a city like Ottawa, where we are developing our software, they are likely to associate dogs with outdoor scenes and leashes, and cats with indoor scenes.

In other countries, there are lots of stray dogs and cats, and household pets are not as popular. So we would predict that if you asked someone from a Canadian city and someone else from rural India to imagine a scene with a cat in it, the Canadian would be more likely to imagine a cat indoors, perhaps on someone's couch, and the Indian to imagine the cat outside (these are my predictions, this has not been experimentally tested).

But we only have the data from the ESP game, which is just images from the internet. Now the images on the internet do not represent what people all over the world see day-to-day super well. So I acknowledge that using pictures from the web as a proxy for human perceptual history has its problems. However, some things are pretty constant all over the world, like the fact that cars are often on roads and that the sky is often above mountains. In any case, we think of our software as simulating the imagination of the typical internet user.

The software is called SOILIE: the "Science of Imagination Laboratory Imagination Engine" (SOILIE). The software gets some input like "dog" and retrieves four objects that correlate with "dog" to put in the picture. In humans, an image of an object takes about twenty to fifty milliseconds to be created in visual working memory, though they might be created in parallel with another image.[9]

So SOILIE is asked to imagine a scene with "dog" in it, and it retrieves four other co-occurring labels. When we did this, however, we noticed a problem.

For example, when we put in "mouse," the words we got back were these: rat, ear, computer, animal, monitor. Notice something fishy? We did, too. When *people* imagine a scene with a mouse in it, they might imagine the animal mouse or a computer mouse, but one thing we know for sure is that they don't imagine some hybrid that has a mixture of both "mouse" meanings!

So my then-master's student Michael Vertolli tried to solve the problem. His solution, in essence, was that SOILIE would get its initial list, and then look at the *global* correlation between all of the elements. As you can imagine, "mouse" correlates with both "rat" and "computer," but "rat" and "computer" don't correlate very well with each other. So SOILIE would replace one of the returned labels with a new one (that was also correlated with the input "mouse") and repeat until all the objects correlated with each other to some acceptable degree. Doing this, he found that SOILIE would naturally converge on one or the other meaning of mouse.[10]

Now, in a human being, the animal "mouse" and the computer "mouse" are two different words, and it is likely that this ambiguity is resolved before the imagination process begins. That said, there are times when the same meaning can result in very different classes of images. Take a birthday party, for example. Some birthday parties have three-year-olds, loot bags, and kid's games, and other birthday parties have bikini-clad women jumping out of cakes. But, unless you hired the world's worst party planner, these elements are not at the same party. So Mike's addition to the software, we believe, might reflect how people resolve these ambiguity problems.

So now SOILIE has a list of things to put in the image. But where do they go? It would be great if the ESP game gave us information about *where* in the image those labels were, but alas, it doesn't. But didn't Luis von Ahn come up with another game to get *that* information? Oh yes, he did. Luis is one of my heroes, because he keeps coming up with such clever ideas.

This next game was called "Peekaboom." As in the ESP game, people are randomly paired up. But in this game, one player is the "boomer" and one is the "peeker." The boomer sees one of the images from the ESP game and is given a label associated with that picture. The peeker sees a blank screen, and it's her job to guess the label that the boomer is looking at. It works like this: the boomer clicks the picture, which reveals a few pixels on the peeker's screen. The boomer clicks more, and more of the picture is revealed to the peeker. After a few clicks, usually, the peeker can see what is being revealed in the image: a tree! As soon as she sees that it's a tree, she types the label in, they both get points, and then move on to the next image.

When people played this game (neither it nor the ESP game are online anymore), the boomers would tend to click on the object in the image that the label described. So now we have the Peekaboom database, which has everything the ESP game database has, and more: it has the *locations* in the images of those labels. So the sky is near the top, roads are near the bottom. Music to my opioid system!

So suppose SOILIE takes "dog" as a suggestion of what to imagine, and retrieves "leash," "man," and "tree." SOILIE then can start to decide where to put these objects in the image. It puts the query label in the center. Imagine you have a blank canvas, and you're going to make a collage on it. You cut out a picture of a dog and place it in the middle of the canvas.

Then it looks at the spatial relationship between "dog" and, say, "man," across the database. We can get an average angle and distance. So "man" tends to be above "dog," because men are taller than dogs.[11]

This is great, because we don't have to tell the software that roads tend to be below cars—SOILIE can figure it out from the data we give it. We think of this database as the software's visual experience.

We think SOILIE simulates, at an information level of abstraction, how people choose what things go into scenes and how they might be placed in those scenes. One drawback is that SOILIE is, at this moment, only two-dimensional. And although visual mental images in human beings might be two-dimensional, the spatial image that is used to build them is likely three-dimensional. We're working on that now.

In the spirit of Luis von Ahn's "serious games," I designed a game to try to get this 3-D information from pictures. I called it "Quanty," and you can play it at quantygame.com free (note that if nobody else is online, you won't be able to play, so if there's nobody online, get a friend to log in at the same time). In it, two players are randomly paired on the net and are both shown the same picture, with two things highlighted in it. Suppose it's a picture of a bedroom scene. The players might be asked, "What is the distance from the lamp to the window?" The players type in their estimate (Five feet? Seven meters?) and the closer their numbers are, the more points they get.

Quanty relies on a fascinating phenomenon called the "wisdom of the crowd." It turns out that if you ask a bunch of people about some numerical magnitude, like the weight of a bull, the number of jelly beans in a jar, or, indeed, how far it is from the lamp to the window, the average response will be quite close to the truth, and usually closer than any one individual's guess. This works better, obviously, for things we have direct experience with. "The crowd" has more difficulty with the weight of a bacterium, or the distance to Alpha Centauri.

If enough people play Quanty, we can get lots of data about the 3-D relationships between objects in the real world, as represented in the photographs. And this data will be accurate, because we can take the average of their responses. Wahida Chowdhury ran and experiment on Quanty and found exactly that (she also found evidence that it was as fun as Tetris, but I have my doubts about that result).[12]

But recently the field of computer vision has gotten good enough that it can have a program look at a still photograph and make a pretty good estimate of the distance each of the objects are from the camera. We are now trying to use these techniques to create a database of three-dimensional scenes so we can modify SOILIE to use a 3-D version of angle and basically work the same way.

What I've told you about SOILIE so far is intended to work *like* human beings work. SOILIE then creates a picture for us to look at, but we don't think that this part has anything to do with *how* people do it.

SOILIE looks at each label, and then pulls from the Peekaboom database pixels that represent that label. So if the label is "tree," it pulls pixels that were boomed in the Peekaboom game during a turn in which someone successfully guessed the word was "tree." It takes a random tree and places it in the image in the right place.[13]

Figure 1. Here is a picture generated by SOILIE, tasked to imagine a scene with a "car" in it. The presence of the sky and road are roughly in the right places.

Figure 1 shows one of the better examples of SOILIE's output. You're looking at computer imagination. Now, looking at this, it's easy to see what's wrong with it. The car is not directly on the road, and the wheel that was added is too big. We can't deal very well with the size of the objects, because it's hard for a computer to tell what the size of an object is in a photo, because the distance from the camera is not known. People have no trouble, usually, knowing how big things are in photos because we have knowledge of how big things are in real life. We have no such database to give computers. Maybe someday the Quanty game will provide that information for us to use. And recent computer vision technology is getting better at determining size from photos. But as of yet these technologies have not been integrated into the software.

Perhaps the most obvious problem with the image is that there is a whole lot of white space. We can tell just by looking at our imaginations that this isn't what happens with humans. To fix this, we added a system that tries to seamlessly put in a background, which can result in delightful images, such as the one in Figure 2.

Figure 2. SOILIE with an added background, seamlessly meshed with the inserted imagined objects.

But again, I want to stress that these outputs are not supposed to represent how people produce the final mental image. We know that people don't store pixel-like information in their long-term memories, and we don't yet have a good explanation of how they turn a symbolic memory into a pixel-like depiction in mental imagery. So these images are just for demonstration purposes. Currently, my laboratory is working on computational neural models of mental imagery, with the hope that these will help shed light on the issue.

Another problem is that the images created are static and many of our imaginings are dynamic, full of moving parts or changing points of view. Although a complete theory of imagination would, indeed, account for the dynamics of imagery, I don't think it's a waste of time trying to figure out how people make static images. We certainly *can* create static images if we want to, and, further, in the human brain motion is processed by the spatial reasoning system, not the visual one.[14] In fact, some people have brain disorders that make them unable to perceive (let alone imagine) motion: when they pour hot chocolate, they have to be careful, because they can't see the cup filling up over time. They just get one static image replaced with another, and in the meantime the hot chocolate might have overflowed. So although we eventually need to account for both spatial and visual imagery, because they are separate systems, they can be studied relatively independently.

MODELING AND PROCEDURE

Just as a scientist who makes a hurricane simulation isn't trying to create the most powerful hurricane or the best hurricane, whatever that might even mean, we're not trying to build SOILIE to be the perfect imaginer. We want it to imitate the human being, flaws and all, so we can better understand what's going on under the hood in our brains. That's what I mean by "modeling" human cognition.

To take another example, let's look at arithmetic. It's trivially easy to get computers to do arithmetic well. We've had calculator watches since the 1980s. What's hard, though, is getting computers to do arithmetic like human beings do! There are lots of cognitive scientists trying to understand how children learn (and don't learn) math and the systematic mistakes they make. Why are children better at adding two of the same numbers (like 6+6) than different numbers (like 6+4)? We come up with theories and build software based on them. If the pattern of performance is the same for the software and the human, it adds support to the theory behind the software.

SOILIE is like that for imagination. We're not trying to just build any old imagination system that works. It has to work like people do. So we read psychology and neuroscience research and make sure the way SOILIE works is plausible, given what we know about the human mind. This, in turn, leads to new ideas and questions about human psychology. That is why we add restraints like the fact that SOILIE only has five objects in each image. A computer can, of course, represent many more.

If we move beyond modeling human beings, though, there has been great progress in what we might consider computer imagination. Computer science has a whole field dedicated to procedural generation, which is essentially the generation of creative designs done by software. Unlike my work in SOILIE, they are not trying to make the computer think like a person does, they're just trying to get software to work as well as it can.

This is particularly useful in computer games, where software sometimes helps create the environment the player character interacts with. There's a great commercial need for this. *World of Warcraft: Cataclysm*, an expansion of the popular video game, took two years to make, and many players consumed all of that content in the first two weeks after release![15]

The snowboarding video game series *SSX* uses procedural generation to make the snowboarding trails. This allows the creators to create three hundred tracks for the same budget of designing ten tracks by hand.[16] This generation is all done before the game is released. When you sit down to play *SSX*, your console is not procedurally generating anything at that moment. But in other games, the map you're using has never been used by anybody else in the history of the universe, because the game software makes it for you on the spot.

An example of this is the popular *Diablo* series of games made by Blizzard, which have procedurally generated (PG) dungeon maps. So when you play *Diablo*, you never know what the layout will be. This helps with making the game replayable.

There is a long history of using PG to create terrain. It often starts by making a random height map, so we know where the mountains will go, then adding embellishments, such as forests and water. A fascinating project called "Uncharted Atlas" makes maps of completely new fantasy worlds, complete with oceans, rivers, mountains, state boundaries, and place names. The place names are also PG, and even sound like they all come from the same imaginary language![17]

PG can also be used to create, or help to create, story. My graduate student Vincent Breault created a PG system that created quest ideas for video games.[18] Tony Veale is a computer scientist who has made a piece of software called Scealextric, which creates characters and plots for stories.[19] In the commercial world, the popular console game *Skyrim* uses PG quests.[20]

For me, there are two pieces of software that represent the more impressive examples of procedural generation.

The first game is *Dwarf Fortress*, a simulation game in which you try to make a band of dwarves flourish into a civilization. It doesn't look very good from the screenshots. It's actually an ASCII game, meaning that the only graphics are made of letters, numbers, and other symbols that you might have in a text file.

Nevertheless, *Dwarf Fortress* is a staggeringly complex world-builder, primarily programmed by a Stanford University math PhD student. It procedurally generates everything from personalities of individual dwarves (Is this one depressed? Does this one appreciate art?) to the gods, to the watersheds of the terrain, to economic patterns.

I can only recommend this game to very hardcore gamers who don't mind a staggeringly long learning phase. But the simulations under the hood are even more staggering—one aerospace engineer familiar with the project said that the simulations in it are more complex than those of jet wing aerodynamic simulations.[21]

Geographically, a single instance of a game of *Dwarf Fortress* simulates 16,000 square miles of land, 250 miles thick. It has varying elevations, mountains, caves, rivers, mineral deposits, and forests. But rather than creating it out of thin air, it simulates geological change over time, including rain, soil erosion, and rain shadows on mountains. A lifelong project, the creators eventually want to put in plate tectonics.

It creates societies and names of places, based on how good or evil the area is and the language of the local species (dwarf, elf, goblin, and so on). Then it simulates history, a week at a time, for 250 years. Towns are created and fall into ruin, kingdoms rise and fall, and all of this is happening before you even begin to play: it just creates the history of the world you're going to play in. It generates personalities of the rulers of nations and uses them to help play out alliances and wars, influenced by food shortages and weather conditions.[22]

Where *Dwarf Fortress* keeps its attention on the simulation and doesn't bother with graphics, *No Man's Sky* is a gorgeously rendered graphical universe of procedurally generated planets, using similar geological simulations that *Dwarf Fortress* uses, but then goes on to create ecosystems of plants and animals (including the sounds they make)—and unlike *Dwarf Fortress*, it creates completely new animals and plants and represents them graphically. You can fly your spaceship from planet to planet and see a whole world that's never been seen (or created) before, interacting with alien species from the imagination of the software.

These are some of the most stunning examples of software imagination. People are generally pretty skeptical about computers being imaginative or creative, and I think that the most impressive feats of creativity are still from human beings. But what's important to appreciate that these systems are already doing some things far beyond what people can do—or at least much, much faster. Worlds that *Dwarf Fortress* or *No Man's Sky* can invent in seconds or minutes might take days or years for a person to do by hand.

I highly recommend watching videos of *Dwarf Fortress* and *No Man's Sky* (in that order—visually, *Dwarf Fortress* is a better as an opening act). Right now these games are viewed on a computer screen, but more and more games are being created for virtual reality systems.

VIRTUAL REALITY

So what is virtual reality? If we think of the real world as providing the input for what we normally perceive as reality, virtual reality is when this input is replaced by something not real. Generally, this means something computer generated. The classic virtual reality setup includes what's called a "head-mounted display," which provides separate little monitors for each of your eyes, providing an immersive, three-dimensional simulation of a world, and blocking input from the real world. In the mid 2010s, these virtual reality "rigs" got much better and cheaper, prompting several companies to create sets that regular consumers could purchase. At the time of this writing, several virtual reality sets are available to consumers, including the HTC Vive, Oculus Rift, and Google Cardboard.

A related concept is augmented reality, which is when you can see the real world, but also computer-generated imagery projected onto it. For example, one might have an augmented reality system that puts text indicating people's names next to their faces when you meet them. This, too, tends to use head-mounted displays, but they are more like augmented glasses, allowing you to see the world as it is, with graphics thrown in. At the time of this writing, emerging augmented reality consumer products include the Microsoft HoloLens and Google Glass.

Any given augmented reality system can vary in how many graphics it has. We can think of augmented reality on a spectrum, where one end of the spectrum is just reality, with no computer stuff at all, and at the other end is a full virtual reality where you get no input from the real world. Because systems can lie anywhere on this spectrum, the whole enterprise is called "mixed reality."

We can see that these technologies mirror what we already do with our own minds. For example, when you close your eyes and imagine being on the beach, you're blocking out reality and trying to replace the input to your visual system with imaginings from your mind. In one case, the computer is replacing reality for sensory input, and in the other, it's your memory. Imagination is a kind of virtual reality, even though people rarely talk about it that way. Similarly, dreaming and full-world hallucinations, such as those that often come with temporal-lobe seizures, are also kinds of virtual reality.[23]

We can also use our imaginations to generate augmented reality. When you look at your living room and try to picture what the wall would look like painted a different color, or how a new chair might look in there, you are projecting your imagination onto the real world, just as a computer might in more traditional augmented reality. Have you ever heard the advice to picture your audience naked to help you calm your nerves when public speaking? This is augmented reality, too.

IMAGINALITY

What a lot of people don't realize is that the imagination is also at work during computer-generated mixed reality experiences. It's not like your imagination shuts itself off when you're in virtual reality. You can picture something that might be behind a virtual wall, for example.

Christopher Stapleton and I suggest that mixed reality, or "virtuality," has three sources, not just two. We can talk about experience being a combination of reality, computer generated content, and the imagination. All three can converge to create the whole of our sensory experience, or what is known as the sensorium.[24]

When we look at computational imagination, we see a great variety in how it's done and why. This helps us keep our mind open to how imagination might work in human beings. From the inkling of wanting to imagine something, to the generation of the imagining, what steps does the mind take?

11

How Imagination Works

The work of science is like that of a composer who can never quite finish his masterpiece symphony. It is a series of drafts that get better and better. At some point the composer realizes it's going in the wrong direction and whole movements get scrapped, and it takes lots more incremental progress to get it as good and then finally better than it was before. The history of imagination research reflects this.

Psychologists' first interests in imagination were, in what we now know as "imagery." The most striking thing about imagery has always been how much it's like perceptual experience. As a result, the questions psychologists asked (in the early 1900s) were the same kinds of questions they were asking about perception: questions of vividness, information capacity, detail, and so on. They were also interested in how these characteristics, as well as things like speed, differed from person to person (what psychology calls "individual differences").

Even early on, some scholars found that people could solve problems without any conscious imagery at all.[1] Although some debated the existence of imageless thought, a movement in psychology called "behaviorism" banished imagery to irrelevancy, at best, and nonexistence, at worst.

For most people, the idea that there is no such thing as mental imagery seems absurd. If there's no such thing as mental imagery, then what is happening when you "see" your dog in a dream? As unintuitive as it might sound, the objections to the idea that mental imagery exists are sophisticated and not easily dismissed—and as we will see, the debate about it continues to rage through philosophy and psychology to this day.

We'll start with the objections of the behaviorists. The behaviorists were a group of psychologists who wanted to make psychology more scientific. Frustrated with Sigmund Freud's legacy and psychology's (then) generally poor experimental methods, the behaviorists sought to clean things up by focusing not on mental entities like beliefs, ideas, and perceptions, but rather on outward behaviors. Importantly, behaviors are more directly measurable, unlike informational entities of the mind.

The benefit of studying behaviors is that they are objectively observable. That is, we don't have to rely on people's reports of what's going on in their heads, we can just measure what they're doing in the real world—the things everybody can see. And indeed, we have countless examples of when people are wrong about what goes on in their heads. So instead of making theories about how, say, beliefs change, they would set up experiments in which people would learn and behave differently—to the behaviorists, a "belief" was not a legitimate scientific concept, because it was essentially unobservable.

A more radical form of behaviorism went even further than that. Rather than just saying that you couldn't run scientific experiments on mental entities, the radical behaviorists said that they didn't even exist. That is, not only can you not study them, but it's irrational to believe that they exist at all! Behaviorism was incredibly influential, especially in the United States and Canada. Europe, not so much.

This all hit research on imagination and imagery very hard. It's not that there were different behaviorist ways of studying them, it's that behaviorism considered such things unstudiable. And for radical behaviorists, the imagination simply *didn't exist.*

Luckily, research in imagery started to blossom in the 1950s and 1960s, for a few reasons. Behaviorism was in decline (and was, for the most part, replaced with cognitive science and cognitive psychology). Research on brainwashing revealed that people were having spontaneous imagery. People were experimenting with drugs, many of which caused hallucinations. Pilots in the war hallucinated planes and other things that were not there. Behaviorism could not come up with learning theories to account for these phenomena, which were getting harder and harder to dismiss as simply nonexistent. For some things, science could not afford to ignore imagery because imagery issues were becoming practical problems.[2]

Currently the psychological debate can be characterized by the disagreements between two prominent psychologists. Stephen Kosslyn believes that mental imagery exists as its own separate representation, and Zenon Pylyshyn (pronounced pill-ISH-in) says it doesn't. This is the imagery debate.

We'll start with what these two agree upon: that long-term declarative memories are represented propositionally. That's a mouthful, so I'll break it down. By declarative memories, we mean memories that are fact-like, such as the memory that your name is Gina or that there is a dog snoring on the couch next to you. This is in contrast with nondeclarative memories, such as your memory of how to ride a bike, how to say the word "milk," and the memories that associate two other memories (for example, that the smell of coffee reminds you of the morning).

What does "propositional" mean? It means that these memories are stored in a sentence-like structure. This does not mean that they're stored in English—they're not stored in any natural language, such as English or Vietnamese—but that they are symbols that are structured together in a way that is similar to how language is. So you might have a memory of your dining room that includes the fact that THE-LAMP IS-ABOVE THE-TABLE. Here, IS-ABOVE is a symbol that connects the concepts of the lamp and the table in a specific way such that the lamp is above the table.

Both Kosslyn and Pylyshyn agree that long-term declarative memories are represented this way. So far so good.

MENTAL IMAGERY

So, our long-term memories aren't pictures, but it *feels* like we can generate pictures in our heads sometimes. Most people claim to experience it this way—that is, they can generate experiences in their minds that resemble what they can see or hear in real life. It's a sensory-like experience. Scientists disagree, however, about what gives rise to these experiences. It certainly *seems* (to most of us) like we are seeing pictures in our minds, but is this an illusion, or are these experiences actually caused by underlying representation and brain processes that are, in some way, picture-like? Some scientists think that not only don't we store pictures in memory, but we don't even generate them when we image something. It only feels like we do.[3]

This distinction might be hard to understand, so I'll try to explain it with an analogy to the way computers work. Some picture formats, like TIFF, describe every point of color, or pixel. Others, like JPEG, also do this this, but in a compressed form. So where a TIFF image might say, "This pixel is black. This next pixel is black. This next pixel is black," a JPEG might say "the next three pixels are black." In both cases, though, they basically represent pixels.

But some formats are completely different, like the saved files of Adobe Illustrator or a computer-aided design (CAD) program. Sometimes called a "vector format," rather than describing the color of points, it describes things like shapes and lines. For all of these, when you view them on a computer screen, it's *turned into dots of color* on the screen, even though the underlying representation, in the file, is quite different.

These two broad ways of representing a picture mirror the debate in psychology: when we experience mental pictures, is the thing in our heads best described as a bunch of points, like a TIFF file, or a more structured representation, like a vector-format file? This is a debate about the nature of the representation of what feels like a picture in the head.

The pictorial theory of mental imagery holds that this is what happens: the sensory-like *experiences* are caused by sensory-like *representations*. What that means is that it represents little more than color points at certain locations. In the extreme version of this view, there is nothing symbolic or meaningful about each point.[4] For visual imagery, it's what is sometimes called "depictive." There is considerable disagreement (and confusion) about what exactly this means. If you are imaging (using your mental imagery to picture) a jar of peanut butter, how exactly is the thing created in your mind picture-like?

Nobody thinks that if we opened up your brain we'd see the color brown in there, visible to our eyes, just as there aren't really pigmented areas in a JPEG image file on your computer. For me, the best interpretation is that there is a matrix of neural tissue, and the neurons in this tissue change their firing rates to represent things like colors and edges of objects. The locations of these things are organized on the tissue much like they are in your eye (the retina part) when you're looking at something in the real world. So, it's not exactly a "picture in the head," even though we sometimes talk about it this way. It's just neurons firing, biologically. It's their *organization* that might be picture-like, creating pixel-like representations.[5]

What would be the function of this? Well, this mental image can then be "looked at." Not with your eyes, but with the same visual attention and perceptual processes you use for light, but instead focused on a mental image. To use the computer analogy again, this might be like a piece of software running through the images you have on your computer's drive, looking for images with your face in them—all without displaying anything onto a monitor.

And this act of perception works like it does when perceiving things in the real world. It recognizes things, creating or recognizing symbols based on the imagined, depictive picture in your head. This is what you probably did when you counted the number of windows in your house. Most people don't know that number by heart, so they need to count them. They make a virtual movie of walking through the house, in their heads, and as they walk through, they "look" at all the windows and keep count. This is what I mean be perceiving a mental image without your eyes. You're using other, later parts of perception.

To take another example, let's have some fun with my initials: imagine a capital letter "D." Now turn it so that the flat side is down. Now put a capital letter "J" beneath it, so that it is touching the "D." What does it look like? Take a moment and try it before reading on.

What's going on here is that you are taking symbols from long-term memory, imaging them, and then using your perceptual abilities to identify a new symbol: many people spontaneously perceive an umbrella from this exercise.

Note that the symbol for "umbrella" was not simply retrieved from memory, nor otherwise previously associated in any important way with these letters. You might already know that the capital letter M looks a bit like an upside-down, capital letter W. You might not need to make a mental picture to figure this out. But with the umbrella example, unless you've done it before, you can't just retrieve the answer: if you're like most people, you create a mental picture in your head, and look at it. *Aha! There's an umbrella!* You've made a discovery using your own imagination. You were able to reperceive the shapes of the letters to identify something new. Reperception is considered to be one of the main functions of imagery.

Now that you have done that exercise with the J and the D (I've done it many times), you now associate the symbol for umbrella with those letters in memory. The next time you do the exercise, you might not need to picture anything at all, because like the visual similarity of W and M, you can simply retrieve the symbol for umbrella. So you've stored "umbrella" to be associated with these letters. But the depiction, the actual pixels you created to extract that symbol, did not get stored in long-term memory. Your mental imagery is like a blackboard that gets erased in less than a second unless your mind keeps refreshing it. Only declarative descriptions are stored for longer periods of time. So, you might create a depiction, and perceive things in it, and those perceptions result in new long-term memories that are propositional.

So how much disagreement is there on the pictorial theory? Ninety-four percent of scholars who study this kind of thing believe that depictive mental imagery can be created in many people's minds.[6] This is the nature of the "imagery debate" in psychology, which has gone on for over thirty years. One of the main problems is that, theoretically, anything we can get people to do in an experimental setting that appears to be the result of using imagery can also be explained by some nonimagery theory.

SPATIAL IMAGERY

So, if the reason we have imagery is so that we can reperceive and reinterpret things, we should be able to get people to reinterpret things in laboratory settings.

There's another famous image, called the "duck/rabbit illusion" that depicts a duck if you imagine the creature facing one direction, and a rabbit if you picture it facing the other direction. When you're looking at it in the real world, you can choose which way to look at it and see it either as a duck or as a rabbit.

Can people who initially see the duck perceive the rabbit later simply by imagining this image? Scientists had people memorize these figures, which they initially saw as either the rabbit or the duck, and then to imagine the figures and see if they could perceive the other animal. It turns out that 80 to 90 percent of people can't do it. You might think this means they didn't memorize it well, but that's not the case—those same people were very capable of drawing what they were imagining, and from there, nearly 100 percent of people could reinterpret the image as the other animal![7]

Here's another example: if I were to show you an outline of the state of Texas, but rotated, so that it's sideways, it will not look at all familiar. One study showed people this and had them study it so that they could imagine it again. Then they were asked to rotate their mental image of it and see if they could tell what it is. It turns out this is very hard to do. Very few people come up with, "Oh, it's Texas!"[8] So if the purpose of mental images is supposed to be reinterpretation, why is it so hard?

Let's review what we know about visual mental images so far: they are generated from symbols in memory and they don't last long unless they are constantly refreshed. When you're trying to reinterpret a mental image, you need to keep refreshing it to give perception a chance to figure something out about it. So people are actually pretty good at reinterpretation, if it does not require manipulating the mental image, like rotating it or resizing it. When you ask people about the shape of a bear's ears or to scan a mental map, it's pretty easy. But rotating a complex shape, like that of Texas, is rather taxing.[9] But there's more to it than that.

To explain what's going on, we need to go one level deeper. Not only is imagery just one kind of imagination, but there are different kinds of imagery that arise just for the kind of stuff you get from light in your eyes! Specifically, there seems to be a difference between visual imagery and spatial imagery. Visual information is about how things look, in terms of shape, color, texture, etc. Spatial information is about where things are in relation to each other—either objects in a scene or parts of a single, complex object. Although when we perceive the world and imagine scenes, these kinds of information feel part of the same experience: we have some good reasons for thinking that the visual and the spatial use different processes in the mind and brain.

When you see an animal in the duck/rabbit illusion, your mind has some idea of which side of the image represents the front of the animal. This information, regarding front and back, is spatial information, not visual. This is why some reinterpretations don't work so well with visual imagery. The duck/rabbit illusion requires a change in which side is the front, which is difficult to do visually—because it's a change in *spatial* understanding.

Let's look at another famous illusion: the "My Wife and My Mother-in-Law" which, depending on how you look at it, looks like either a young or an old woman.

This illusion is much easier to reinterpret, when you ask people to do it later, with visual imagery (as opposed to spatial), because it requires reinterpreting parts of the image, without moving it, flipping it, or changing which side is front (all spatial transformations).[10]

To help understand what spatial information is, think about what a blind person might imagine. If someone's been blind since birth, a "congenitally blind" person, they have no experience with visual information and report no visual mental imagery.[11] But, crucially, they have what we consider *spatial* information from walking around, touching things, and from acoustic information from the reflection of the sounds in rooms. Spatial imagery is three-dimensional, unlike visual imagery.[12] They certainly get a sense of where everything is—how do they do it?

They do use imagery, but do not use *visual* imagery. Think about kicking off your shoes under the dinner table. At the end of the meal, you might feel around for them and put them back on, all without looking. In this case, you are using memory and sensation based on touch and the feeling of where your body parts are. And when you imagine where your shoes are, you might not experience it as a visual mental picture. Blind people can have images that are not visual but rather spatial, proprioceptive, and based on touch. Spatial knowledge can come from any of the senses and can also come from understanding texts, as when someone verbally describes to you the layout of their home.

Congenitally blind people appear to be able to do tasks that would require *spatial* imagery, but not specifically *visual* imagery. Visual and spatial tasks seem to use different brain areas, and brain damage to different parts of the brain can impair one and not the other.[13] Engaging in visual perception interferes more with visual imagery than spatial imagery, and engaging in spatial or motor tasks interferes more with spatial imagery.[14]

Another way to think about spatial imagery is with this exercise. Look up from your book and imagine, as vividly as you can, a woman standing in front of you, wearing a very bright red dress. Now imagine that she walks around you so that now she's standing behind you. In your imagination, don't change the point of view—that is, don't imagine her behind you as though the "camera" were behind you, and you're in the picture—keep "looking" forward.

What most people experience is that when the woman goes beyond their visual field (that is, outside of where they can actually see), the image changes in character. The color is gone, and it's more like *knowing* that she's there, but not really *seeing* that she's there. This, for many, is the difference between visual and spatial imagery.[15]

I do have to say that this doesn't work for everybody. People who don't experience visual imagery see no change, because it's spatial imagery no matter where they imagine her being. But some people I talk to claim to have 3-D *visual* imagery, complete with color, all around them. Research has found similar results.[16] I don't know what sense to make of this, because my understanding is that the visual imagery area is roughly the size of the visual field, and you should not be able to visually image things outside of it. But maybe some people can.

Unfortunately, lots of tasks that scientists assume require visual imagery can actually be done with either spatial or motor imagery. For example, when (congenitally) blind people are asked whether a capital letter A has any curved lines in it, they often imagine drawing the letter to find out, thereby doing the task using motor rather than spatial or visual imagery!

As I've mentioned, when a person is required to change an image to reinterpret it—either by rotating it, changing which side is the front, and so on—this seems to be a spatial imagery transformation, not a visual one.[17] But one task appears to be truly visual: imagining foreshortening. For example, if you imagine a plank, with one end close to you and the other end a few feet away, understanding that the other end of the plank looks smaller (in terms of the area it takes up in your visual field) seems to be a purely visual task. There's not much from touch or sound that suggests that faraway things look smaller, and, as we'd expect, congenitally blind people are quite bad at inferring that foreshortening happens at all (though foreshortening can be figured out from echolocation).[18]

As I discussed earlier, lots of people experience mental imagery, such as pictures in the head, but not all scientists agree that this means there are depictive representations that give rise to them. So, there's a difference between the experience and the underlying processes and representations that gives rise to it. Many scholars, such as Zenon Pylyshyn and Peter Slezak, think that *nobody* has underlying depictive representations. They think that people's *experience* of mental imagery is an illusion—that people think they have depictions in their heads, but they really don't. Opponents of the depictive theory believe that the experience of mental imagery is generated only from propositional representations. That is, you can have the experience without the depictive representation.

I believe that we have depictive representations in our heads, and for the rest of the book I'll assume that we do, but I just wanted to make it clear that not all scholars agree on this matter.

CONSCIOUSNESS AND UNCONSCIOUS IMAGERY

Now when most scholars think about this issue, they assume that if someone has a visual image then they would be conscious of it. But *might* it be possible to have the depictive representation without the experience? When people experience a visual mental image (the experience of pictures or sounds in the head), it's the awareness of this underlying depictive representation. But could it be that some people have depictive visual representations, *but don't know it?*[19]

If so, then some, or maybe all, people with aphantasia (the inability to *experience* mental imagery) actually do have imagery, but don't know it. Maybe we have three classes of people: those who have no pictures in their head and (obviously) can't experience them, those who have them and can't experience them, and those who have them *and* can experience them, which is most people.[20]

If this sounds hard to believe, I'll try to make it sound more plausible by telling you about a real example that's even harder to believe: a phenomenon called "blindsight." In this condition, patients think they are blind (or blind in a part of their visual field). That is, they have no conscious experience of what their eyes are taking in. In their brains, what's happening is that the signal from the eyes is not reaching their cerebral cortex. However, the eyes send information to other parts of the brain that apparently doesn't result in much conscious awareness. As such, patients claim they can't see anything, but can guess pretty accurately about properties of what's in front of them.[21] They can see, but think they're blind. (There is the opposite disease, Anton-Babinski syndrome, in which patients really are blind but believe they can see!) So, if people can see without knowing it, it seems reasonable that they might be able to create and use depictive representations and not know it.

So what can people with blindsight see? They can often pick out visual features, such as edges or motion, but don't have any grasp of a holistic visual percept (almost everything about blindsight, I should mention, is controversial in science).[22]

As I was writing this book, a man emailed me with an interesting description of his experience. He is aphantasic in all senses—he has no conscious experience of imagery in vision, hearing, or anything else. He explained to me the relationship between his imagination and emotions. Sometimes we use imagery so we can feel a certain way. For example, if I want to cheer up, I can imagine playing with a bunch of puppies. I might close my eyes and image it as vividly as possible, everything from what the puppies look like, to the sound of their little yips and snorts, and the feeling of their fur on my fingers and their little tongues licking my hand. When I do this, I get happy. Or, you might visualize winning some prize to get the feeling of pride.

But this man has a very different experience. When he tries to imagine something emotionally evocative, it just doesn't work. It's a conceptual imagining—what happened, who was there—but it has no emotional punch. It's clearest in sexual imagery. Like many people, sometimes he tries to imagine sexual situations to generate sexual arousal in himself. But he can't get turned on by his conceptual imagination.

Here's where it gets weird: when he *tries* to use imagery, that is, he tries to make a mental picture rather than just imagining the situation conceptually, he gets the emotional change (that is, he gets sexually turned on) even though he experiences no imagery.

So, if he imagines a sexual situation without trying to use imagery, it doesn't feel like anything. He just has a cold, thoughtful understanding of the imagined situation. But when he tries to picture it, to hear the sounds of the imagined sexual situation, the feelings come, but no sensory-like experience accompanies it. This is only one case, but it is suggestive that this man has subconscious imagery. What might be happening is that he creates a sensory image of the sexual situation, which affects his emotional state, yet somehow this image never makes it through to consciousness.

It's not just for sex, either. He gets the same effect when he tries to relive the good feelings associated with getting praise for a job well-done or getting scared in a frightening situation.

I asked some of my other aphantasic friends about this, and one of them had no trouble getting emotional responses from conceptual imagination, and the other didn't get any and couldn't even understand what it even meant to "try to generate imagery." Any emotions she'd get from memories were involuntary. So here we have three reports, and none of them agree with each other. Welcome to psychology. Make yourself comfortable.

Clearly we need more science on this, but unconscious imagery is a terribly understudied idea, for a few reasons: the very idea of unconscious imagery is a little hard to wrap your mind around, the idea that some people don't experience imagery is only just now starting to be well known, let alone the idea that some people might not experience imagery but unconsciously still have it. Lots of psychologists don't even appreciate the difference between spatial and visual imagery.

For example, studies have been done that measure how much of a visualizer a person is, and also how well they do at mental imagery tasks. A classic one is called "mental rotation" in which a participant looks at two objects and tries to determine if they are the same object, in two different orientations. In one of best results in cognitive psychology, Roger Shepard found that people tend to take more time when one of the objects would need to be rotated more to match the other one—about 60 degrees per second. Crucially, later studies found that visualizers took the same amount of time as nonvisualizers! Critics of depictive mental imagery presented these findings to show that depictive imagery representation (pictures in the head) was a bogus idea. But we now think that this rotation task is a spatial task, not a visual one, so we wouldn't expect vividness to correlate with rotation abilities (a study by Amedeo D'Angiulli found that it does, however).[23] So perhaps people can have unconscious *spatial* imagery.

There are many studies that suggest that many nonimagers seem to perform equally well on many tasks that are assumed to require imagery.[24] But many of these studies were done without a good distinction between visual and spatial imagery. That is, the scientist might have assumed that the task required visual imagery, but in fact they were mistaken, and the task could have been accomplished using only spatial imagery.

Okay, but what about unconscious *visual* imagery? Can anybody's mind generate a visual image but not be aware of it? To find evidence that someone had unconscious visual imagery they would have to be able to do a task that *required* visual imagery—that is, it could not be accomplished by spatial imagery or imagery in any other sensory modality. If unconscious imagery were possible, we would predict that certain tasks that seem to require, or are benefited by using imagery, are done with the same skill levels no matter how much imagery a person tends to experience. If you lose the ability to generate mental images, then you can't be conscious of them. Such people would perform poorly on visual imagery tasks. But people who can create them but just don't know it should still perform pretty well on them—the only deficit we should see would be for those tasks where the conscious awareness was actually an important part of the process.

Unfortunately, this kind of experiment is difficult to set up.

So it seems that whether or not they can have unconscious visual imagery remains an open question, in terms of rigorous science.

I suspect that many people picked this book up because they care about imagination and would like to hone their skills with it. Given the usefulness of imagination in so many endeavors in life, this is a laudable goal. There's no evidence yet that aphantasics can improve their experience of imagery with practice. But what about the rest of us?

What's disturbing is that spatial reasoning abilities start to decline after you are twenty years of age. This is true even for people who are in jobs that require visualization, such as architecture—architects have *better* visualization than people in fields that do not require visualization, but they still experience an age-related decline.[25]

You can make your mental imagery more vivid, in the moment, by lying down. We don't know why, but people have speculated that it has to do with the relaxation of neck muscles, associations with sleep and dreaming, or simply because ceilings don't have much on them to distract us.[26]

IMPROVING IMAGERY AND IMAGINATION

But can you improve your general ability to use mental imagery? There are some promising studies that suggest that, like many things, improving your ability to have a vivid imagination is a matter of practice.

As far back as 1883, Sir Francis Galton reported being able to improve people's ability to use mental imagery through practice and by sketching their images in the air. Over the course of a few months, they were able to make good drawings of the things they practiced imaging.[27] A study by geologist Sarah Titus showed that students' visualization skills improved after taking a course in geology—a field that requires being able to think about the structure of Earth in three dimensions.[28] David Ben Chaim ran a study in which he had about a thousand kids build and draw structures made of cubes. They showed improved imagination abilities even a year later. But, for this study, the test of imagination involved questions about arrangements of cubes—the exact domain they had training with.[29] These studies seem to improve "skills," which are different from vividness and detail in imagery. Also, they are limited to a particular domain. That is, practicing visualization in geology helps with geology problems, and practicing visualizing cubes helps you with cube problems, but do we have evidence that your visualization skills improve, in general?

A few things suggest that imagery, *in general,* can be improved. María José Pérez-Fabello found that a few years of art training improved people's visual mental imagery abilities, and an old study by Carol Parrott showed that imagery training can improve vividness.[30]

A very interesting episode of the podcast *Radiolab* describes two men who went blind but adapted in very different ways. John Hull decided that the visual world was no longer relevant to his life. When he went blind, he stopped engaging in mental imagery and, instead, focused on the senses that he still had. By choice, he no longer has visual mental imagery. Zoltan Torey, on the other hand, made a deliberate effort to visualize everything. He would make up pictures in his head to match the other stimulation he was getting from the world.[31] Zoltan's imagery life was vivid and full, suggesting that practice (or the lack of it) can affect your general imagery abilities.

We've also seen evidence of imagination training from the religious perspective. People have to do intense practice of imagery before they start to spontaneously hallucinate religious revelations.

An experiment by Joanna Nobbe found that if stroke victims practiced using mental imagery twice per week for six consecutive weeks, they experienced a 22 percent increase in vividness, but this study only had seventeen participants, and the statistics were barely significant.[32]

Another paper was supposed to have evidence, but all copies of it seem to have been lost in the flood of Brisbane.[33]

REASON FOR SKEPTICISM

So as far as careful scientific study goes, this isn't very much evidence at all, although there are plenty of fascinating anecdotes. Unfortunately, the field of general imagery improvement is big on advice and small on good science. Many books and articles cite "evidence" that, when you look into it, are mere assertions with weak or no science to back them up.[34] For example, the "blackboard technique" involves imagining writing the letters of the alphabet on a chalkboard—A, B, C, etc. Every time you write a letter, you try to keep the image of all of the previous letters in mind (at the time of this writing, I can only do A through E). With practice, your mental imagery is supposed to improve, and you will be able to image more and more letters. But I couldn't find any evidence that it works. And even if it did, it might only apply to imagining letters.[35]

And there are good studies that *failed* to find an effect of mental imagery training. Rosanne Rademaker used a clever experiment to test the effectiveness of mental imagery training that did not need to rely on subjective reports of vividness. Rather, her experiment used a known bias in perception that is affected by imagery. She had people engage in visual imagery for an hour a day for five days, but found that there was no increase in the expected effects of imagery. Also, counter to the other studies I mentioned earlier, she found no increase in reported vividness.[36] Another study, by Alan Richardson, had people imagine food and looked to see if vivid imagers had more salivation, or if training in imagery would increase salivation. The findings were small to nonexistent and short-lasting.[37] It is unknown whether people who have no imagery abilities at all can do anything to change their situation.

What are we to make of this? First of all, there are far too few studies to know conclusively whether or not general imagery vividness or detail can be improved with training or practice. But my gut tells me that it doesn't work or, at most, it has small effects. My reason for thinking this is based on the "file-drawer" problem. Although Rademaker's study was published in a good journal, most studies that fail to find effects also fail to get published. Reviewers and journal editors don't like them, and consequently scientists often don't even bother to write them up or submit them—writing up a study for publication takes a lot of work, so scientists don't want to bother if they expect the paper will be rejected anyway. This results in lots of studies that scientists do getting shoved in the metaphorical file drawer, never to be seen by other scientists or the public. The dearth of good results in the literature might be because lots of people have tried to find effects of vividness training and failed to find any, and then moved on to other research projects without ever publishing their poor results. Not very promising.

Rademaker has a suggestion for why she failed to find an effect—because mental imagery happens at lower-level visual states closely related to vision. Perhaps these brain areas are difficult to change, either because they are more hardwired for our visual needs, or perhaps so well-trained by our environment that the minor amount of training used in a laboratory experiment isn't enough to budge it. Or, as worded in her paper, "Imagery may simply lack sufficient impact to induce permanent plastic changes at these lowest sensory levels."

What is hopeful about this interpretation, though, is that it leaves open the possibility that higher-level imagination, which happens in the more conceptual areas of the brain, might still be able to be changed through practice and training. With hope, future research will shed light on this possibility.

So what are we to make of the tulpamancers, fiction authors, people with imaginary companions, and religious people who learn to create hallucinations as a result of training? My colleague Michael Vertolli suggested a solution to this apparent paradox: perhaps these practices make you better at creating autonomous *characters* (through automatization) but do not affect your basic imagery abilities, such as vividness and detail, which we might not be able to change.

At this point in the book we've seen imagination in its many forms. Mental imagery, dreaming, hallucination, creativity, planning, and so on. Now we're ready to take stock and talk about the space of what imagination can be. Below is a list of all of the ways that one imagining can differ from another. I'll describe each one.

WAYS IMAGINATIONS DIFFER

- Voluntary or involuntary
- Images appear full-formed or are assembled bit by bit
- Personal significance: meaningful or meaningless
    - Triggers a search for explanation as to a cause
    - Reacts to hallucinator or doesn't (is the imaginer a participant or spectator)
- Insight: believed or not believed
- Sensory (which senses), delusory (nonsensory meaning), or both
- Simple or complex (patterns vs. objects/animals/people)
    - For audiovisual scenario only: linguistic complexity (just words, full sentences, or conversations)
- Scale of vividness—or loudness, if auditory
- Scope of detail
    - Localization: projected onto external space, (like augmented reality)
    - Just in the visual field (like an afterimage)
        - Or more akin to virtual realty-like
        - Or just "inside the head"
- Conscious or unconscious
- Repetition (systematized/repetitive)
- If it is voices in the head: gender, familiarity

Now We Turn to Each Subset:

VOLUNTARY OR INVOLUNTARY

Can the imagining be brought about by an act of will? Voluntary imaginings are brought about by a conscious effort. The voluntary and involuntary distinction was first introduced by sociologist Philippe Bûchez with respect to hallucinations. He considered only the involuntary kind to be pathological.[38]

What would it mean for a hallucination to be voluntary? One example I found in the scientific literature is the study in which people were asked to imagine hearing a song, and, subsequently, some of them actually believed that it was actually playing. Here we have a voluntary image that, somehow, people lost insight on. I suppose this counts as a voluntary hallucination, although the people in the study did not intend for the image to be a hallucination, merely auditory imagery.

More convincing is a report of a man who suffered from auditory verbal hallucinations after being tortured in war. He would experience hearing previously heard conversations and claimed to be able to induce them with an act of will.[39]

There are strategies that people use to control their hallucinations once they happen, including trying to change the features of them, stopping them once they start, developing ways to cope with them, and attending to how they change under stress.[40] But it seems that for most people with hallucination symptoms, they cannot simply will them never to happen.

Normal imagination is often voluntary, but we can also experience spontaneous imagination that is not experienced as hallucination—presumably because we maintain insight that it's just an imagining.[41] We can choose to imagine a jar of peanut butter, but we can also get unbidden images brought to consciousness. This can happen when people have PTSD and cannot stop thinking about their traumas, but this can also happen when someone tells you a story and you can't help but image it. Sometimes you might feel an image coming to your head, and you can choose to make an effort of will to stop it. This might be seen as halfway on the voluntary-involuntary spectrum.

PERSONAL SIGNIFICANCE AND MEANINGFULNESS

Of particular importance to hallucinations is whether they are personally significant or meaningful. The visual hallucinations of Charles Bonnet syndrome, for example, are notable in that they are meaningless and carry no emotional punch. In contrast, many schizophrenic hallucinations are drenched in meaning. They feel important, likely because they are unusual and come with strong emotional charges, which trigger the person to search for personally meaningful reasons and causes for what they think they're seeing. This can lead to delusions or other strange beliefs about the world.[42] Voices heard might seem powerful, possibly even omnipotent, and the hallucinator might feel compelled to follow the voices' commands. These kinds of beliefs about voices in the head correlate with how much distress comes with hearing them. Interestingly, people who have negative views on the voices in their heads also tend to have negative views of real people, perhaps learned from life experiences. It might be that their distress comes from their negative views on others, be they real or imagined.[43]

Some software models of perception and hallucination suggest that hallucinations occur when there is high uncertainty, which provides an explanation for why sensory deprivation can cause hallucination, but also might shed light on why hallucinations can often be paranoid—social situations are much more nuanced than many other perceptual situations. Seeing that the peanut butter is brown, for example, is much simpler than being able to tell if someone is getting annoyed with you. It could be that the inherent uncertainty involved with perceiving others' emotional states and motivations makes paranoid hallucinations more common.[44]

It's easy to think that the meaning and personal significance of a hallucination, dream, or other imagining is based on the content of it. That is, you feel fear if it's a scary vision. But as we've seen in dreams, it might be that the imagining, in general, can be a response to a feeling that comes first. It's theoretically possible. You also might have experienced people who are anxious, and whatever topic you bring up, they find some reason to worry about it! This, too, seems to be the emotion driving the imagination.

But, of course, it works in the other direction, too, as when you might use a visualization exercise to relax. In these cases, the imagination comes first, and your emotions are a response to it.

INSIGHT

Insight, or transparency,[45] is your ability to know that you're imagining or hallucinating, rather than concluding that your experience is of something in the real world. This is not an all-or-nothing category: insight might be delayed, partial, or even fluctuating.[46] Typically, we associate a lack of insight (that is, being fooled by the contents of your imagination) with dreams and hallucinations, but as we've seen, even asking people to imagine hearing music makes some of them think they're really hearing it—effectively losing insight they previously had. Recall, also, that there are many cases of hallucination in which at first there is no insight, but insight later comes. When one first gets a ringing in one's ear (a very simple hallucination), one might at first think it's in the environment, but quickly concludes that it's all in one's mind, based on, for example, how it doesn't change from room to room or with changes in head orientation. We can even get insight when dreaming, as in the case of lucid dreams.

SIMPLE OR COMPLEX

Imaginings also vary in complexity. The early visual system has detectors for things like lines and simple shapes. These can often be hallucinated with migraine auras, certain drugs, or even just by rubbing your eyelids. The specific shapes, lines, spirals, zigzags, and so on, appear in simple hallucinations, and also tend to appear in pictures created by our ancestors in the Upper Paleolithic.[47] These are sometimes known as phosphenes. This suggests that some art has the shape that it does because of the nature of our early visual system. Perhaps it is created by people with these simple hallucinations, and they are recording what they hallucinated. Or perhaps seeing these in paint or print feels very compelling because it's a fundamental part of our perceptual wiring. I think both things are probably happening.

In many sensory modalities, we can get high complexity, corresponding to the higher-level perceptual areas in the brain. Typical voluntary imagination and schizophrenic hallucinations might have something close to the detail and complexity of perceiving things in the real world.

We see variation in complexity in verbal hallucinations, too. Hallucinated speech might be just single words, or sentences, or full conversations. Further, the hallucinator might just hear it, talk back to it, or converse with it. These conversations might be between the hallucinator and the hallucinated voice or might be between two hallucinated characters in the hallucinator's mind.[48] Although I've seen no data on it, it's interesting that imagined (nonhallucinated) speech tends to be complex sentences and conversations rather than individual words.

VIVIDNESS AND ACOUSTIC CLARITY

Vividness is the strength, or intensity, of the imagining. For voluntary imagination, the vividness is under voluntary control, although each person has their own upper limit (which decreases with age) of how vivid their imagery can be. Although hallucinations can also vary in their vividness, the vividness of your imagination, interestingly, does not correlate with your likelihood of suffering from hallucination.[49]

Voices in your head, be they hallucinated or simply imagined, might sound like real voices, with gender, loudness, clarity, and so on, but they also might just be like verbal thoughts, with no particular acoustic properties.[50]

Interestingly, vividness is different from detail.

DETAIL

Detail refers to how much information is in a mental image or hallucination. Not only is this different from vividness, but there seems to even be a trade-off: people with a bigger visual cortex (area V1) have more detail but less intense images! It's as though the size of V1 determines how much detail there can be, but there is a constant amount of imagination energy, and when there is a lot of detail, this energy is spread out, resulting in a lower overall intensity. People who have schizophrenic and PTSD hallucinations, which are known for their intensity, have smaller V1 volume and fewer neurons there.[51]

LOCALIZATION

Another factor characterizes *where* the imagining is imagined to be. This is the imagining's localization. For visual imaginings, one might have one's eyes open and imagine a jar of peanut butter on the floor. Because these imaginings are projected by your mind into the environment, I will call them "projected" imaginings. Suppose you hallucinate, or imagine, a jar of peanut butter on the table in front of you. If you turn your head and look away, so that the table is not in your visual field, you still will have a conceptual and spatial imagining of where the jar is, but any visual imagery will go away, only to return when you orient your head back to the table, where you "put" it.

Sometimes simple hallucinations will stay in one place in your visual field, but not in your environment, such as when you look at a bright light and then get an afterimage. I call these "retinal" imaginings. It seems to be right there on your retina. Like a speck on your camera lens, you can't simply look away from it. Anywhere you look, the image follows.

Dreaming, and some hallucinations, are "full world." This means that one's immediate sensory environment vanishes or becomes irrelevant, and the person has an experience of being in another place entirely. This isn't projected into the environment, because the imagining is detached from your sensory reality. Nor is it meaningfully anchored to the retina. It's a full virtual reality created by the mind. When I'm meditating, sometimes I go to a serene place—for me it's a screened-in porch by a river. This is a full-world imagination. I close my eyes and try not to attend to my sensory environment. Full-world imaginings are a kind of localization, too.

There's a fourth kind—an imagining that seems to have no location at all. On several occasions, when giving a talk on imagination, I ask people to imagine a cat. They don't imagine a full scene with a cat, and they don't imagine the cat projected into their actual environment and don't particularly associate their mental image with any place on their retina, either—the location is simply indeterminate. This is possible because the what and where pathways of the visual system are separate, and you can have a sensation (real or imagined) without an idea of where it is.

So, localization of visual imagination can take one of these forms: projected, retinal, full world, and location-less.

This distinction is also applicable to sound imagination. Sometimes when people experience hallucinated sounds, they are projected. The hallucinations appear to come from a particular spot—in the next room, above you, or something like that. Sound hallucinations might seem like they're coming from your immediate environment, but also might seem to be coming from outside of your normal hearing range—such as outer space.[52] Other times it just sounds like it's a voice in your head. (Or, more rarely, coming from another body part!) Some of these voices sound like you, and sometimes they sound like somebody else. It is likely that these issues arise from problems with the areas of the brain involved with sound localization and distinguishing the self from others.[53]

CONSCIOUSNESS

In a broad definition of imagination, which would include any new mental construction, it's pretty clear that we could have imaginings that we are not conscious of. The existence of unconscious mental imagery, however, remains controversial. The idea of an unconscious hallucination seems to make no sense—anything not experienced wouldn't be categorized as a hallucination in the first place.

REPETITION

Imaginings can be the same thing over and over, or each one can be idiosyncratic.[54] For voluntary imagination, this is not particularly important—one might have a repeating sexual fantasy, for example. But for disorders it can be diagnostic. For example, someone who has been traumatized might have repeated mental reenactment of the event or a recurring nightmare. People with hallucinations sometimes have the same hallucination over and over. For repetitive hallucinations, psychologists might also care about how many times the repetition happens.

THE PROCESS OF IMAGINATION

So now we know the space of possibilities of imagination, and we're ready to talk about the process of how imaginings are generated—how does it all work?

The elements of your imagination come from your memory. Indeed, where else would they come from? Although the prime candidate for the constituents of imagination are your episodic memories—that is, memories of what you've experienced yourself—your semantic memories are important, too. These are memories that are not tied to any particular experience—such as what your name is or what number comes after five.[55]

QUERIES AND PROMPTS

When we think about imagination as a process, it has a beginning and an ending (and then, possibly, a loop). The beginning is whatever makes you start to imagine in the first place. People imagine for all sorts of reasons. I will list several, but there are likely many more we have not thought of.

We might be asked directly to imagine something, as I have done several times in this book: Imagine a triangle. Imagine a jar of peanut butter. Imagine a complicated anteater. Imagine a four-sided triangle.

We also use our imaginations to flesh out verbal descriptions, such as when we listen to a story someone is telling us. When we read a novel, and we get a description of a person or a scene, we use our imagination to make sense of it. Imagination is important for the basic comprehension of language—even if there is no conscious mental imagery.[56]

When we read about scenes, we form memories that allow us to reason about them.[57] We certainly create structured descriptions that describe the visual and spatial aspects of what we're reading about, though what we have been calling "imagery" might only happen when pictures are also presented, or readers read the text over and over.[58]

CHOOSING FROM OUR MEMORY

What kinds of things can we imagine, and what is impossible to imagine? Whenever I hear a movie trailer talk about something "beyond your imagination," I get a little annoyed. My imagination is quite good, thank you very much!

Let's start with visual imagery. Some things are just too abstract to create an image of. Take the concept of justice, for example. We can imagine a scene of someone getting justice, or being denied justice, but we cannot form a picture of justice itself, on paper or in our heads—certainly not one that would be easily recognized by others. We can image something that symbolizes justice, such as Superman or a courtroom, but those images better describe what they literally depict: Superman and a courtroom.

Even concepts like "furniture" are hard to image. Furniture includes tables, chairs, and sometimes big speakers. There is no single image that captures the idea (that wouldn't be better described by another word). The same goes for food. When I travel, I like to look at signs trying to communicate in pictures that there is no eating or drinking allowed. Often, the drink is represented as what appears to be the silhouette of a soft drink container one might get at a fast food restaurant. The "food" is often a hamburger, or French fries. But it's clear that not all food looks like either of these. Sometimes, when I walk my dog, I see signs indicating that there are no dogs allowed. The dog pictured looks, to me, like a miniature schnauzer. My dog's a pug, so maybe he's welcome?

Another interesting aspect of these signs is that they have to represent that something is forbidden. We tend to use the "no symbol" for this, which is a circle with a backslash through it, often printed in red. But negation, or prohibition, doesn't really look like anything, and it only means "no" because we have a cultural agreement about the symbol's meaning, kind like how we know that "&" means "and." Someone unfamiliar with this symbol would not recognize what it means, because it doesn't look like anything in the natural world.

The situation is very different for concrete, physical objects and their parts. Psychologists have recognized that there is a "basic level" of categorization. Things at this level include shoes, shoelaces, dogs, and hammers. The things at this level share many properties. For example, when we consider all "hammers," we can see that we would interact with just about all of them in basically the same way. We cannot say the same thing about the class of "tools." Basic level things are quickest to identify and are learned first by children.[59]

Most relevant to imagination is that the basic level is the highest level of categorization that we can make a picture of (and still be able to recognize what it is). All hammers look kind of alike, but we can't say the same for all tools.

So when we are imagining something, such as a scene, we might imagine an abstract concept, such as clothing or justice, but when we actually go to create a mental image, we are going to have to choose some concrete object that can be pictured. You can *imagine* buying a generic piece of furniture, but to *image* it, you have to pick a particular kind of furniture, such as a chair. Mental images are more concrete and less abstract.

OBJECT PLACEMENT AND ANIMATION

Once an object has been selected, it must be placed somewhere in the scene, relative to the other objects. Our spatial memory of where things tend to be informs these unconscious decisions. For instance, you know that there's usually a road or driveway beneath a car and that dining room tables might be empty or might have plates and utensils on them. Again, this is something your mind decides without your being conscious of it. When you imagine a car accident, you don't have to consciously put the cars on roads, they just seem to appear out of nowhere.

We don't only have prototypes of objects, such as dogs or bananas, but also of whole scenes, such as birthday parties and farms. These scene prototypes can provide a spatial structure to inform where objects should go.

As we have seen, even remembering something from your own experience involves a reconstruction that resembles the processes used in imagination. When recalling a scene, the general structure is retrieved, but then further details are only fleshed out when those details are attended to.[60]

MIRRORING, VISUAL MEMORY, AND PERCEPTUAL HISTORY

The elements of our imagination come from our memory, and if we focus on imagery, they come from what we might call our perceptual history—what you've seen and heard your whole life influences how you imagine things. Let's take a simple example: imagine a triangle that's filled with exactly your favorite color. Try to see it in your mind as clearly as you can, with the color as vivid as you possible.[61]

Unless you're trying to be particularly creative, you probably imagined a roughly equilateral triangle with the flat side down—most people do. Let's examine why.

The equilateral triangle is the kind of triangle we see the most often. If you do an internet image search for "triangle," you'll see what I mean. A triangle that is one inch wide and a mile high is still a triangle, but it's not the kind of triangle you see very often. So the simple explanation is that we imagine triangles like we do because they reflect our perceptual history—we see them in our mind's eye the same way we usually see them with our eye's eye. We can think of this triangle as the prototype. Prototypes of all kinds, be they of animals or mathematical concepts, are processed faster, and they tend to pop into consciousness first.[62]

But we can dig deeper. Why are most of the triangles we see in the world like this? Why did culture converge on the flat-side down version?

Well when we see triangular-shaped things in the real world, such as the sides of pyramids, we often see them with the flat side down because of gravity. The flat side down is more common because it's more stable for real objects in a world with gravity. Place a triangular object on its point and it will probably fall over so that a flat side is down. That's also why we think of rectangles and squares as having a flat side down. In fact, if a square has a corner down we use a different word for it altogether: diamond.

SUPPORT

Locations of things in our imaginings are also constrained by how things are supported by gravity. For example, we not only know that a lamp might be above a dining room table, but we would also know that it must be supported somehow—by hanging from the ceiling, or otherwise kept from falling. Likewise, the plates are technically "above" the table, but they also must be touching the table, not floating a few inches above it. These examples might seem obvious, but when you are modeling imagination on computers, they become big deals and allow us to appreciate the processing our brains must be doing. We've learned by computer modeling that a simple distance and angle isn't enough to know the relative location of imagined objects, because objects need to be supported by something, even if it's the ground.

METAPHOR

However, there are lots of decisions our minds have to make that don't have anything to do with these things. There is a large body of evidence that we think of abstract things metaphorically. When we think about abstract things like love or minds, we think of them, metaphorically, in terms of more concrete things like journeys and containers.[63] These metaphors likely constrain our imaginations as well.

One metaphor that has a lot of evidence is GOOD IS UP (traditionally, metaphors are written in all capitals), which is reflected in our language ("He's feeling down") and in our religion, where the divine is often described as being above us. It's possible that if we are going to imagine two things, we might put the one that we think is more "good" above the one we think is more "bad."

This has not been directly tested, but other experiments are suggestive. Daniel Richardson had people look at English verbs and then choose which direction they were going in. They would hear a verb, such as "respect," and then choose which direction (up, down, left, or right) felt the most appropriate. This sounds like a strange idea for an experiment, I know. We all know that in the real world "respecting" someone has no direction. It's a matter of how you act, and what you say, things like that. But in this experiment, you force people to just pick a direction. The interesting thing is that people tended to agree on verb "direction." It should not surprise you that "respected" is an upward direction verb, and negative words like "smashed" are downward verbs. Given that these verbs don't really have direction, the theory is that these results are due to metaphorical associations in people's minds—in this case, probably GOOD IS UP.[64]

Similarly, we have a left-to-right bias. That is, people tend to perceive motion moving from left to right in their visual field as being better than right to left. This is reflected in films, where the protagonist is often moving left-to-right on the screen. When people are asked to draw a circle pushing a square, they tend to draw the motion left-to-right. But preschoolers don't have this bias, suggesting that it's cultural.

Further, people whose primary language is Arabic have the opposite bias, because that language is written right-to-left. Even the Hebrew version of FedEx has a logo with an arrow going the opposite direction of the English FedEx logo. These data support the theory that the left-to-right bias we have is due to the direction of writing.[65]

Why might this be? We can think of our personal space as being defined by three axes: front-back, left-right, and up-down. Of these, only the up-down axis is asymmetric in the physical world—that is, because of gravity, up is very different from down. Things that are up tend to fall down. Things that are higher up tend to be smaller. Living things start down and grow up. Healthy, awake, and energetic people are higher up than the sick, sleeping, and dead people. Because the world is like this, the theory goes, we have a very deep association with up and upward motion being "good," and our language, gestures, diagrams, and imaginations reflect that. Nearly all diagrams of geological ages and of evolution, for example, put human beings and the present day at the top.[66]

Think of world maps, and how north is almost always displayed as up. Why might this be? There's nothing inherently "up" about north; it's just convention. But the cultures that came to dominate the world (and thus made the maps that are used internationally)—European and the Northeast Asian countries—happened to be located in the northern hemisphere. Assuming an egocentric bias, perhaps those ancient cultures put themselves on top and everything else below, because it conformed to their deep association between "up" and "good." In English, we have the expression "to go south" when something gets worse. If Australian, South American, or African nations had dominated the world at this period in history of peak cartography, our maps might have been created south-side up.

The front-back axis is only asymmetric relative to our own perspective. In general, we can see and act on things in front of us, but not for things behind us. The left-right axis is weakest and seems only to be affected by being right- or left-handed and writing direction.[67]

How should this affect our imagination? When we think of people doing something, we will be likely to imagine the doer on the left and the direction of action on the right—at least for "good" behaviors. To my knowledge, this has not been tested directly.

PERCEPTUAL ANTICIPATION

So far in this chapter I've mostly been talking about imaginings that consist of symbolic structures. But what about the sensory quality that many imaginings have? What happens when we experience mental images in our head? How are they created? The top theory is that mental imagery is a little like running your perception system in reverse.

In perception, a person takes in energy or chemicals from the world (light, sound waves, or chemicals in food or the air) and turns them into patterns of firing neurons. This is essentially what you do with your eyes, nasal passages, tongue, fingertips, and ears. In the eyes, for example, there are cells that fire more rapidly in the presence of light or when light changes intensity. From there the sensory systems analyze this neural signal, figuring out what's in it.

For vision, we talk about low-level and high-level vision. Light entering the eyes gets processed by the vision system first for what we call low-level properties, such as the edges of objects, surfaces, things like that. At the higher level, memory needs to get involved, because you need memories of how things look to be able to recognize them. (Low and high to talk about visual processing? Remember the GOOD IS UP metaphor!) So the early visual system can largely work without accessing long-term memories, but later visual systems use memory.

To help speed up the process, the system anticipates what is expected. This makes complex vision possible. Suppose you hear your dog walking toward you. You might turn to look. Your visual system is expecting the dog, and in the process of preparing your mind to quickly see the dog, the parts of the visual system that would recognize the dog become readied. Technically, the neurons become more active, as though you were already seeing the dog you expect. Because the neurons are primed, you can recognize your dog more quickly. Anticipation makes your perceptual system more efficient. Sometimes this is called "predictive coding."

One theory of imagery is that this perceptual anticipation is the process that leads to mental imagery and hallucination. In essence, it suggests that your mind "expects" something so much that the same parts of the perceptual system are so highly activated that it's like you're really seeing it. This is not to say that every time you imagine something, it's because you expected to see it, but rather that imagery uses the same processing elements in your mind that are used in perceptual anticipation.[68]

If this theory is right, though, it poses a bit of a mystery regarding why some people experience mental imagery and others do not. Most people's vision is intact, so those anticipatory parts of the visual system are working, so if imagery uses the same parts, why can't they be used for imagery? If visual imagery is the same as expectation, we would expect that aphantasics would be worse at those visual tasks that require anticipatory expectation. To my knowledge this has not been studied, but aphantasics seem not to be visually impaired.

RENDERING IMAGERY

In visual perception, we go from simple detections in the light patterns to complex perceptions, such as seeing someone devouring a jar of peanut butter. In imagination, we are going in the opposite direction: we might start with the idea of someone eating a jar of peanut butter, build a scene description of it, and then, sometimes, make a mental image. Just as in perception, the early system detects simple visual features, and the later visual system detects more complex arrangements, every time someone uses mental imagery they are starting from ideas and pushing it toward the eyes. The further this process goes, the more perceptual the experience of imagination is. Eventually this might result in "depictions": pixel-like images composed of dots of color (and intensity and depth) at particular locations, probably represented in the visual buffer in the occipital lobe (spatial images, as they are already represented topographically in another part of in the brain, don't need to be rendered in the visual buffer).[69] Again, I want to stress that there is currently an "imagery debate" happening in psychology that argues the very existence of mental imagery in this final form.

WHERE THE MAGIC HAPPENS

The visual buffer is where this depiction is theorized to appear, generally located in the primary visual cortex part of the brain's occipital lobe.[70] If this part of the brain represents mental images as points of color in some two-dimensional space, it makes sense to ask about the "resolution" of this buffer, as though it were a computer screen. The resolution of this buffer is something equal to or less than the number of neurons in it. It might be that several neurons are used to make a single pixel, but there probably aren't more pixels than neurons. It doesn't matter, though, because there are *millions* of neurons in this region of the brain.

The dots can be thought of as being in some two-dimensional coordinate space. Polar coordinates support easy rotation and scaling,[71] but we don't really know what the brain does exactly, but it's probably some kind of coordinate system that is used for spatial reasoning.

How far "down," toward the eyes, can imagination go? The most well-known imagery researcher, Stephen Kosslyn, believes that even for people with very vivid mental images, the activation only goes to the visual buffer (in the occipital lobe). There are a few old studies that suggest that for some, the activation can go all the way to the retinal cells (some people seem to experience complementary color afterimages after vivid imagining), but most scientists I've talked to about it are skeptical of this.[72]

As we have seen, engaging in visual and spatial imagination resembles vision, but going backward through the system. This is fairly accurate as a rough idea. But visual imagery goes no further than the visual buffer in the occipital lobe. This means that the early visual processing, such as that done by the neurons in your eyes, cannot be affected by imagination.[73]

REINTERPRETING IMAGERY

The theorized function of imagery is the reinterpretation of something you've experienced before. You might look at a car, and then a friend asks you to look again because the front looks like a face. You look back, and are able to re-perceive it, and see the face. The idea is that with mental imagery, you can do that with your own visual memories.

When you mentally walk through your childhood home, counting the windows, you are doing reinterpretation, because the number of windows was not previously associated with those images. Right now my wife and I would love to put a piano in our house. It's hard, though, because we are not sure where we could fit one. When we reason about this, we use our imaginations. How big is the piano? Could someone get one up the stairs? How would it look in the guest room? We've seen pianos before, and we know our own house really well, but we need to use our imaginations in this situation, because we have not had experience with a piano in the house before. That's why we can't simply retrieve from memory where a piano would fit—we have to simulate it in our imaginations.

Unfortunately, getting people to reinterpret mental images under laboratory conditions has proven difficult. Some studies find that people can do reinterpretation, others find we have a lot of trouble with it.[74] When people imagine the famous duck/rabbit illusion, which looks like a rabbit facing right or a duck facing left, it is rare for people who initially interpreted it one way are able to reinterpret it in the other way using their mental imagery. For visual imagery, it seems that reinterpreting low-level visual features is pretty easy, but if you have to do any kind of "reconstrual," such as flipping front and back, rotating, etc., it becomes a spatial task, not a visual one, and this explains why some reinterpretation tasks are easy and some are hard.[75] That is, it's easy to get visual properties from a visual mental image, but a visual mental image is difficult to manipulate with spatial transformations.

So if reinterpreting the duck/rabbit would involve spatial but not visual imagery, this might explain why people can't do it with visual imagery very well. But in these experiments we're not asking people to use only visual imagery. We're just asking them to do the task—indeed, the participants have no idea that there is a difference between visual and spatial imagery anyway. So if they can do spatial imagery, then why can't they solve the duck/rabbit task in their imaginations using it? When we ask people to image the duck, why do we assume that their image is only visual, and that they are not simultaneously creating a spatial image that is manipulatable? To my knowledge, this question has not been answered.

I want to reiterate that pure mental imagery—depictive representation that consists of little more than colors at locations—is only one kind of representation for imagination. When you generate an imagining, the mind makes hundreds of decisions, and likely generates many structured representations at various levels, before finally "rendering" a 2-D image in the visual cortex that we'd call a mental image.

Thinking along these lines, we can look at metal imagery as one extreme on a continuous scale from abstraction to specificity. If I ask you to imagine a star shape, for example, you would retrieve from memory your concept for the star shape from the phrase "star shape" that you read. This symbol is very abstract. From there, you would create the basic structure of the shape, perhaps in terms of lines, or corners, or intersections—some structured description, perhaps with annotations for orientation, color, line thickness, whether it's filled in or not, the location, anything else in the scene with it, and so on. What happens from there is still mysterious. You might generate a series of what vision scientists call "edges," which are where color changes happen in an image. Finally, possibly, you get down to the lowest level, which is a series of representations of colored dots in your visual cortex—the mental image. So, for the image of a star, there would be a bunch of neurons that are firing in a way that represent the existence of some colors at some locations, and these described points would form the shape of a star.

What this means is that if you come up with a clear visual mental image of a star, your mind is actually representing it at many levels of abstraction at the same time.

FADING

Having created a visual mental image of a star, without continuing to attend to it, the picture will fade quickly. This is particularly noticeable in imagined scenes with lots of objects. Visual mental images are complex, and the image creation process, though it might be fast, is not instantaneous, and it is built piecemeal, with different parts showing up in the mental image at different times. But because this happens in the visual buffer, which fades rapidly, what we have are different parts of the image being updated and fading at different rates.[76]

You might experience this when you try to picture a complicated scene. As your attention moves around the mental image, new details are formed, and the parts you're not attending to become more indistinct. When you attend to those areas again, your mind refreshes that part of the image. To keep a mental image in place, it has to be constantly refreshed from other parts of memory—it's a very, very short-term store. The reason for this, as I've discussed above, is that the mental image is being formed in the same part of the brain that is used for vision, and if it didn't fade, we'd experience something like smearing of what we're seeing now with what we saw a few milliseconds ago. Another ramification of this is that the more complicated the image is, the more parts will have faded by the time the last element of the picture has been added.[77]

The fading is faster for smaller imagined objects. This is likely due to the nature of that part of the brain. The connections between neurons are mostly inhibitory—that is, when they fire, they slow the firing rate of other neurons, rather than speeding them up. The connections are also short, because a given neuron is most closely connected to the neurons near it, because they represent nearby space. As a result of this, small areas of the visual buffer will enjoy a great deal of inhibitory activity, usually more than can be generated by rest of the brain to keep the image active. Remember, mental images are made from memories, and the other parts of the brain have to constantly pump information into the visual buffer to keep the mental image in mind. When the mental image is larger, there are fewer inhibitory connections between one part of the image and another (because the neurons are distant and more unconnected), so it takes less input to keep the mental image in mind. This is why smaller images are harder to maintain. It's not a problem with normal vision, though (we don't experience more fading when we look at smaller things), because the activation from the eyes is strong and constant, and requires no effort.[78]

Remember that the perceptual systems are multilayered, and imagery corresponds to the lower, more sensory levels. The higher-level detectors are likely to have a longer persistence.[79] This has been found with both vision and with hearing. So while low-level visual images don't maintain themselves, the higher-level, more conceptual visual and spatial ideas might persist without as much constant refreshing.

For most of us, most of the time, mental images are not as vivid as perceptual vision, and this corresponds to brain studies that show that the visual buffer is less active during imagery than it is for vision.

CONCLUSION

By now you have read about what might seem to be a bewildering number of visual and spatial things in human brains. Sorry, but that's the way it seems to be. We have an iconic memory, which seems to store raw, mostly uninterpreted information from the eyes. This is stored only for a few hundred milliseconds. We have visual working memory, which is visual or spatial information from long-term memory put in short-term storage so that our minds can work with it. Further, we seem to have depictive mental representations in the visual buffer, for mental imagery, that can come from perception or from long-term memory in the form of mental imagery. Its effects on perception last a few seconds, which is how we know it's a separate system from the short-term store in iconic memory.[80] Further, we have different systems for spatial and visual perception and imagination (and imagery).

A few scientists have suggested that human beings are the only animals who can use imagination. Certainly we are the best at it. But there are studies that suggest that, depending on how you define imagination, other animals do, too.

Adam Johnson ran studies in which rats ran through mazes. He analyzed their brains and found that when the rat is deciding on which path to take, its brain creates a representation of one path it might choose, and then another. Interestingly, these path representations represent only the path *in front of* the rat, not behind it, suggesting that the rat brain considers possible futures before making a choice. We don't know whether the rat is conscious of these deliberations, but it doesn't matter—if we consider imagination to be a recreation or generation of a new representation of the world, distinct from the current environment, then these rats are, indeed, imagining.[81]

But even with studies of the limited imagination abilities of animals, such as crows and rats, there's no doubt that no other animal can use their imagination with abilities anywhere near that of human beings. We can imagine complex possible futures, fantasy worlds, fantasies about being on a beach, the mundane and the majestic. We can use our imaginations to make us relaxed—or anxious. The most impressive feat of human imagination, to me, is our ability to use it in creative endeavors. Sitting in a chair, with our eyes closed, we can imagine what the world might be, and construct elaborate plans.

With such power, we have an obligation to use it for good—to make the world better for ourselves, and for the world. So imagine how the world might be better, in both big and small ways.

Then go make it happen.

Final Words

If you review this book online, please consider using the sentence "it was better than peanut butter" in the review. You will be in on a joke with other people who actually finished the book.

Thank you my agent, Don Fehr, and to Jessica Case and the team at Pegasus.

Thank you to Michael Vertolli for collecting references and reviewing the motor imagery research, and to Lia Turner, who did extensive background research on hallucinations.

Thank you for all of the students who participated in Carleton University's Science of Imagination Laboratory, including Eve-Marie Blouin-Hudon, Kae Bagg, Mackenzie Ostler, Can Mekik, Jack Parsons, Judy Hoang, Amy Cheng, Tianxue Zhang, Joanna Kapron, Vincent Breault, Jon Gagne, Ivy Blackmore, Sebastien Ouellet, Michael Cichonski, Sean Riley, Justin Singer, and Dr. Sterling Somers.

This work also benefitted from conversations, online and off, with Jardri Renault, Jennifer Groh, Jeanette Bicknell, Myrto Mylopolous, Amedeo D'Angiulli, Steven Franconeri, Maithlee Kunda, Robert West, Steven Lindsay, Peter Slezak, Daniel Reisberg, Frank Tong, Renaud Jardri, Tobias Egerton, Gilles Fénelon, Jane Ransom, Daniel Schacter, Daniel Reisberg, Elizabeth Marsh, Ann Taves, Gregory Schankland, Chris Timmermanns, Darren McKee, Kim Hellemans, Bill Faw, Patrick Orr, Taylor Howarth, and Shawna Tregunna. I want to thank all of my friends on Facebook. I ask them a lot of questions and get great answers.

Thanks also for the support from my family, James and Janet Davies, my sister, novelist JD Spero, and my beloved wife, Vanessa Davies.

The most thanks go to my former academic advisors Dorrit Billman, Nancy J. Nersessian, and Ashok K. Goel. These three spent an enormous amount of effort to train me to be a scientist, and for that, I will be forever grateful.

Endnotes

1. IMAGINATION: WHAT IT IS

1 Richardson, A. (1983). Imagery: Definition and Types. In A. A. Sheikh (Ed.) *Imagery: Current Theory, Research, and Application.* New York: Wiley and Sons. pp. 3–42, 15.

2 Philosopher Shaun Nichols refers to conceptual imagination as suppositional imagination, or "p-imagination," and creating perception-like mental states as enactment imagination, or "e-imagination."

Nichols, S. (Ed.). (2006). *The Architecture of the Imagination: New essays on pretence, possibility, and fiction.* Oxford: Oxford University Press. pp. 41–42.

3 Dennett, D. C. (2013). *Intuition Pumps and Other Tools for Thinking.* New York: W. W. Norton. p. 290.

4 Wilkinson, G. S. (1988). "Reciprocal Altruism in Bats and Other Mammals." *Ethology and Sociobiology*, 9(2–4), pp. 85–100.

5 Harari, Y. (2014). *Sapiens: A Brief History of Humankind.* New York: Random House. p. 26.

6 This argument is beautifully articulated in: Harari, Y. (2014). *Sapiens: A Brief History of Humankind.* New York: Random House.

7 Psychologist Larry Barsalou disagrees, and thinks that all memories are stored as perception-like structures. His theory is controversial: Barsalou, L. W. (1999). "Perceptual Symbol Systems." *Behavioral and Brain Sciences*, 22, pp. 577–609.

8 To read more about how we know people don't have photographic memories, see: Foer, J. (2006). "Kaavya Syndrome." *Slate*, April 27. retrieved May 19, 2017 from http://www.slate.com/articles/health_and_science/science/2006/04/kaavya_syndrome.html.

9 Amit, E., Algom, D., and Trope, Y. (2009). "Distance-Dependent Processing of Pictures and Words." *Journal of Experimental Psychology: General*, 138(3), pp. 400–415.

10 Amit, E., Wakslak, C., and Trope, Y. (2013). "How Do People Communicate with Others? The Effect of Psychological Distance on Preference for Visual and Verbal Means of Communication." *Personality and Social Psychology Bulletin*, 39, pp. 43–56.

11 Boroditsky, L., and Ramscar, M. (2002). The roles of body and mind in abstract thought. *Psychological Science*, 13(2), pp. 185–189.

12 Anderson, M. L. (2016). "Précis of after Phrenology: Neural Reuse and the Interactive Brain." *Behavioral and Brain Sciences*, 39, e120.

13 There are esteemed scientists who disagree. Barrett, L. F. (2017). "Emotional Intelligence Needs a Rewrite. *Nautilus*, July/August, 21, pp. 69–73. http://nautil.us/issue/51/limits/emotional-intelligence-needs-a-rewrite.
14 Blouin-Hudon, E. M. C., and Pychyl, T. A. (2015). "Experiencing the Temporally Extended Self: Initial Support for the Role of Affective States, Vivid Mental Imagery, and Future Self-Continuity in the Prediction of Academic Procrastination." *Personality and Individual Differences*, 86, pp. 50–56.
15 A diary study showed that most images people have during the day are visual. Kosslyn, S. M., Seger, C., Pani, J. R., and Hillger, L. A. (1990). "When Is Imagery Used in Everyday Life? A Diary Study." *Journal of Mental Imagery*, 14, pp. 131–152.
16 Jabr, F. (2014). "Speak for Yourself." *Scientific American Mind*. January/February, pp. 46–51.
17 Sacks, O. (2012). *Hallucinations*. New York: Alfred A. Knopf. p. 45.
18 Sommerville, J. A., and Decety, J. (2006). "Weaving the Fabric of Social Interaction: Articulating Developmental Psychology and Cognitive Neuroscience in the Domain of Motor Cognition." *Psychonomic Bulletin & Review*, 13(2), pp. 179–200.
19 Jeannerod, M. (2003). "The Mechanism of Self-Recognition in Humans." *Behavioural Brain Research*, 142, pp. 1–15.
20 Vividness seems to be related to visual but not spatial imagery. Someone with very vivid visual imagery is better at most visual tasks, but it won't tell you anything about how well they will do at spatial tasks, which is further evidence that these are different processes. Reisberg, D., and Heuer, F. (2005). "Visuospatial images." In P. Shah and A. Miyake (Eds.). *The Cambridge Handbook of Visuospatial Thinking*. Cambridge, UK: Cambridge University Press. pp. 62–63.
21 Zeman, A. (2016). "Aphantasia: 10,000 People Make Contact over Visual Imagery. *The Exeter Blog*. Posted November 8; retrieved May 2, 2017 from https://blogs.exeter.ac.uk/exeterblog/blog/2016/11/08/aphantasia-10000-people-make-contact-over-visual-imagery/.
22 Pincus and Sheikh claim that imagery is at the center of consciousness: Pincus, D., and Sheikh, A. A. (2009). *Imagery for Pain relief: A Scientifically Grounded Guidebook for Clinicians*. New York: Routledge. p. 59.
23 This idea, of demonstrating visual imagery by asking someone how many windows are in their home, was created by Jerome L. Singer. Cited in Lazarus, A. (1977). *In the Mind's Eye*. New York: Rawson Associates. p. 4.

Other ways to evoke imagery: ask someone which are longer, the flippers or the feet of a penguin, or which syllable of the song "Happy Birthday to You" is sung at the point of the highest note of the song. These two examples are from Gilbert, D. (2006). *Stumbling on Happiness*. Toronto: Vintage Canada. pp. 129–130.

24 Reisberg, D., and Heuer, F. (2005). "Visuospatial Images." In P. Shah and A. Miyake (Eds.). *The Cambridge Handbook of Visuospatial Thinking*. Cambridge, UK: Cambridge University Press. p. 37.
25 Faw, B. (2009). "Conflicting Intuitions May be Based on Differing Abilities: Evidence from Mental Imaging Research." *Journal of Consciousness Studies*, 16(4), pp. 45–68.
26 Zeman, A., Dewar, M., and Della Sala, S. (2015). "Lives without Imagery—Congenital Aphantasia." *Cortex*, 73, pp. 378–80.

This condition of not experiencing mental imagery is called "aphantasia," which can be congenital (from birth) or the result of surgery or injury: Zeman, A., Dewar, M., and Della Sala, S. (2015). "Lives without Imagery—Congenital Aphantasia." *Cortex*, 73, pp. 378–80.

27 I'll get into the "depictive imagery" debate later in the book.

Kosslyn, S. M., Thompson, W. L., and Ganis, G. (2006). *The Case for Mental Imagery*. Oxford: Oxford University Press. pp. 179.

28 Grinnell, D. (2016). "Blind in the Mind." *New Scientist*, 34, pp. 36–37.

29 Faw, B. (2009). "Conflicting Intuitions May be Based on Differing Abilities: Evidence from Mental Imaging Research." *Journal of Consciousness Studies*, 16(4), pp. 45–68.

30 Sutherland, M. E., Harrell, J. P., and Isaacs, C. (1987). "The Stability of Individual Differences in Imagery Ability." *Journal of Mental Imagery*, 11(1), pp. 97–104.

31 Control Happens in the Frontal Areas. Faw, B. (2009). "Conflicting Intuitions May be Based on Differing Abilities: Evidence from Mental Imaging Research." *Journal of Consciousness Studies*, 16(4), pp. 45–68.

2. PERCEPTION AND MEMORY

1 Bear with me here. The jaunty melody of "MMMBop" hides poignant, sad lyrics that sound much more like they were written by world-weary adults than the teenagers who actually wrote them.

2 It's further complicated by the fact that the same information sometimes involves different complexes of neurons for different people—we all learn things in ways specific to us and have learned different things. We all have our own cognitive histories. This is called "multiple realization" or "multiple instantiation." To the extent that this is true, we *can't* talk about the biology of thought in a way that applies to people in general. So we talk about information moving through the system, which we assume is more uniform from person to person than their exact neural organizations.

3 The next layer of cells, the retinal ganglion cells, are about a hundred times fewer than the photoreceptors, so a lot of information compression has to happen in the eye, before it even gets to the brain.

4 First the information from the eyes goes through a part of the thalamus called the lateral geniculate nucleus and from there goes to V1.

5 Tootell, R. B., Silverman, M. S., Switkes, E., and De Valois, R. L. (1982). "Deoxyglucose Analysis of Retinotopic Organization in Primate Striate Cortex." *Science*, 218(4575), p. 902904.

Technically, these neurons need not be close together for this to work. They only need to be wired up correctly. Just like you can have a computer on one side of the room connected to a monitor on the far side of the room with a long cord, it doesn't really matter where the computer is, or whether it's upside down, whether the cord is twisted, or whatever. What matters is how it's connected. It's the same with this part of the brain. It just so happens that not only is it wired up to represent continuous space, but it's actually laid out like that in the brain: neurons that represent close places in space are actually physically close together in the brain. Because they are laid out in the same pattern as they are on the retina, they are called "retinotopic."

But if all we need is it to wired up so that it works, why does the brain go the extra mile to put the neurons that only need to *represent* closeness in space are also *actually physically* close in the brain? It could be because connecting distant neurons is metabolically expensive. Just like it might cost you more money to buy a long cord to connect a monitor to a computer across the room, it costs human brains nutrition, time, and room to have wiring needlessly going all over the place. So maybe having these neurons be contiguous was simply the least expensive way for evolution to wire them up, because they required shorter connections.

Kosslyn, S. M., Thompson, W. L., and Ganis, G. (2006). *The Case for Mental Imagery*. Oxford: Oxford University Press. p. 22.

6 Levine, D. N., Warach, J., and Farah, M. (1985). "Two Visual Systems in Mental Imagery Dissociation of 'What' and 'Where' in Imagery Disorders Due to Bilateral Posterior Cerebral Lesions." *Neurology*, 35(7), pp. 1010–1018.

7 A more extreme kind of filling in can happen with light and sound. Tommy Kot brought people in the laboratory and repeatedly played them a tone and flashed a light at the same time. After they'd associated these two things, they reported hearing the tone when only the light was presented—showing that a sound hallucination can be created in the laboratory simply through association.

Kot, T., and Serper, M. R. (2002). "Increased Susceptibility to Auditory Sensory Conditioning in Hallucinating Schizophrenics: A Preliminary Investigation." *The Journal of Nervous and Mental Disease*, 190, pp. 282–288.

8 Tversky, B. (2005). "Functional Significance of Visuospatial Representations." In P. Shah and A. Miyake (Eds.). *The Cambridge Handbook of Visuospatial Thinking.* Cambridge, UK: Cambridge University Press. p. 21.

Davies, J. (2014). *Riveted: The Science of Why Jokes Make Us Laugh, Movies Make Us Cry, and Religion Makes Us Feel One with the Universe.* New York: Palgrave Macmillan.

9 Krishna Rao, H. P. (1923). *The Psychology of Music.* Guluvias Printing Works.

10 Wright, A. A., Santiago, H. C., Sands, S. F., Kendrick, D. F., an Cook, R. G. (1985). "Memory Processing of Serial Lists by Pigeons, Monkeys, and People." *Science*, 228(4710), 287–289.

Think of white noise—the sound that comes out of a television on channel fuzz, or a radio when it's not tuned into a station. Any given three seconds of white noise sounds basically the same. But from an information theory perspective, they are all incredibly different. We just don't have the symbols we need to perceive and remember that information.

11 Tarr, M. J., and Black, M. J. (1994). "A Computational and Evolutionary Perspective on the Role of Representation in Vision." *CVGIP: Image Understanding*, 60(1), pp. 65–73.

12 Choi, S., and Bowerman, M. (1991). "Learning to Express Motion Events in English and Korean: The Influence of Language-Specific Lexicalization Patterns." *Cognition*, 41, pp. 83–121.

13 Hayward, W. G., and Tarr, M. J. (1995). "Spatial Language and Spatial Representation." *Cognition*, 55, pp. 39–84.

14 Rosch, E. (1975). "Cognitive Representation of Semantic Categories." *Journal of Experimental Psychology*, 104, pp. 192–233.

15 Tversky, B. (2005). "Functional Significance of Visuospatial Representations." In P. Shah and A. Miyake (Eds.). *The Cambridge Handbook of Visuospatial Thinking.* Cambridge, UK: Cambridge University Press. pp. 1–34.

16 Specifically, it's the parietal cortex: Tversky, B. (2005). "Functional Significance of Visuospatial Representations." In P. Shah and A. Miyake (Eds.). *The Cambridge Handbook of Visuospatial Thinking.* Cambridge, UK: Cambridge University Press. p. 4.

17 Turke-Browne, N. B., Jungé, J. A., and Scholl, B. J. (2005). "The Automaticity of Visual Statistical Learning." *Journal of Experimental Psychology: General*, 134(4), pp. 552–564.

18 The theory that everything should be equally accessible in memory is called the "equiavailability theory." The theory that we recall things with an image is the "imagery theory," and the axis theory is the "spatial framework theory." Tversky, B. (2005). "Functional Significance of Visuospatial Representations." In P. Shah and A. Miyake (Eds.). *The Cambridge Handbook of Visuospatial Thinking.* Cambridge, UK: Cambridge University Press. p. 7.

19 Huttenlocher, J., Hedges, L. V., and Duncan, S. (1991). "Categories and Particulars: Prototype Effects in Estimating Spatial Location." *Psychological Review*, 98(3), pp. 352–376.

20 Franklin, N., Tversky, B., and Coon, V. (1992). "Switching Points of View in Spatial Mental Models. *Memory & Cognition*, 20(5), pp. 507–518.

21 Rinck, M. (2005). "Spatial Situation Models." In P. Shah and A. Miyake (Eds.). *The Cambridge Handbook of Visuospatial Thinking.* Cambridge, UK: Cambridge University Press. pp. 334–382.

In one study, participants seemed to be using categorical ideas of space (number of rooms in between, ignoring size of the rooms) when estimating distance, but Euclidean ideas of space when doing relative distance judgments!

Rinck, M., Hähnel, A., Bower, G. H., and Glowalla, U. (1997). "The Metrics of Spatial Situation Models." *Journal of Experimental Psychology: Learning, Memory, and Cognition*, 23(3), pp. 622.

22 Tversky, B. (2005). "Functional Significance of Visuospatial Representations." In P. Shah and A. Miyake (Eds.). *The Cambridge Handbook of Visuospatial Thinking.* Cambridge, UK: Cambridge University Press. p. 13.

23 This isn't just for places we know well. When people are asked to take the perspective of a faraway place, they overestimate the distances within that faraway place as though they were there. For example, if they imagine themselves in Italy, they will overestimate the distance between Rome and Florence: Tversky, B. (2005). "Functional Significance of Visuospatial Representations." In P. Shah and A. Miyake (Eds.). *The Cambridge Handbook of Visuospatial Thinking.* Cambridge, UK: Cambridge University Press. p. 13.

24 Tversky, B. (2005). "Functional Significance of Visuospatial Representations." In P. Shah and A. Miyake (Eds.). *The Cambridge Handbook of Visuospatial Thinking.* Cambridge, UK: Cambridge University Press. p. 20.

25 Rinck, M. (2005). "Spatial Situation Models." In P. Shah and A. Miyake (Eds.). *The Cambridge Handbook of Visuospatial Thinking.* Cambridge, UK: Cambridge University Press. pp. 334–382.

26 Rinck, M. (2005). "Spatial Situation Models." In P. Shah and A. Miyake (Eds.). *The Cambridge Handbook of Visuospatial Thinking.* Cambridge, UK: Cambridge University Press. pp. 334–382.

27 Tarr, M.J., and Bulthoff, H. H. (1995). "Is Human Object Recognition Better Described by Geon Structural Descriptions or by Multiple Views?" Comment on Beiderman and Gerhardstein (1993). *Journal of Experimental Psychology: Human Perception and Performance*, 21(6), pp. 1494–1505.

These findings are controversial. See: Biederman, I., and Bar, M. (1999). "One-Shot Viewpoint Invariance in Matching Novel Objects." *Vision Research*, 39(17), pp. 2885–2899.

28 See Tommy Edison's YouTube video "What Do Blind People See?" and Davies, J. (2014). "What Do Blind People Actually See?" *Nautilus* August 14 blog entry.

29 Gopnik, A., and Graf, P. (1998). "Knowing How You Know: Young Children's Ability to Identify and Remember the Sources of Their Beliefs." *Child Development*, 59, pp. 1366-1371.

30 Atance, C. M., and Meltzoff, A. N. (2005). "My Future Self: Young Children's Ability to Anticipate and Explain Future States." *Cognitive Development*, 20, pp. 341-361.

31 Schacter, D. L., and Addis, D. R. (2007). "The Cognitive Neuroscience of Constructive Memory: Remembering the Past and Imagining the Future." *Philosophical Transactions of the Royal Society of London: B*, 362, pp. 773–786.

32 Gilbert, D. (2006). *Stumbling on happiness.* Toronto: Vintage Canada. pp. 87 and 112.

33 Gilbert, D. (2006). *Stumbling on happiness.* Toronto: Vintage Canada. p. 88.

34 Brainerd, C. J., and Reyna, V. F. (2005). *The Science of False Memory.* Oxford: Oxford University Press. pp. 370–372.

35 Studies suggest between 81–93 percent of therapists who hear of repressed childhood trauma believe that the events described by their clients actually happened. Loftus, E. (1993). "The Reality of Repressed Memories." *American Psychologist,* 48(5), pp. 518–537.

36 Garry, M., Manning, C. G., Loftus, E. F., and Sherman, S. J. (1996). "Imagination Inflation: Imagining a Childhood Event Inflates Confidence that it Occurred." *Psychonomic Bulletin & Review,* 3(2), pp. 208–214.

37 Reber, R., Schwarz, N., and Winkielman, P. (2004). "Processing Fluency and Aesthetic Pleasure: Is Beauty in the Perceiver's Processing Experience?" *Personality and Social Psychology Review,* 8(4), pp. 364–382.

Whittlesea, B. W. A. (1993). "Illusions of Familiarity." *Journal of Experimental Psychology: Learning, Memory, and Cognition,* 19(6), pp. 1235–1253.

38 Garry, M., and Polaschek, D. L. L. (2000). "Imagination and Memory." *Current Directions in Psychological Science, 9,* pp. 6–10.

39 Malmquist, C. P. (1986). "Children Who Witness Parental Murder: Post-Traumatic Aspects", *Journal of the American Academy of Child Psychiatry,* 25, pp. 320–325.

40 Loftus, E. (2003). "Make-Believe Memories." *American Psychologist,* November 58(11), pp. 867–873.

Reisberg, D., and Heuer, F. (2005). "Visuospatial Images." In P. Shah and A. Miyake (Eds.). *The Cambridge Handbook of Visuospatial Thinking.* Cambridge, UK: Cambridge University Press. pp. 66 and 68.

41 Brainerd, C. J., and Reyna, V. F. (2005). *The Science of False Memory.* Oxford: Oxford University Press. pp. 225—223.

42 Holmes, E. A., James, E. L., Coode-Bate, T., and Deeprose, C. (2009). "Can Playing the Computer Game 'Tetris' Reduce the Build-Up of Flashbacks for Trauma? A Proposal from Cognitive Science." *PloS One,* 4(1), p. e4153.

43 Rusting, C. L., and Nolen-Hoeksema, S. (1998). "Regulating Responses to Anger: Effects of Rumination and Distraction on Angry Mood." *Journal of Personality and Social Psychology,* 74(3), p. 790.

44 Lindsay, S. (2008). "Source Monitoring." In H. L. Roediger, III (Ed.), *Cognitive Psychology of Memory. Vol. 2 of Learning and Memory: A Comprehensive Reference,* 4 vols. (J. Byrne Editor). Oxford: Elsevier. pp. 325–348.

45 Reisberg, D., and Heuer, F. (2005). "Visuospatial Images." In P. Shah and A. Miyake (Eds.). *The Cambridge Handbook of Visuospatial Thinking.* Cambridge, UK: Cambridge University Press. p. 66.

46 Johnson, M. K., Hashtroudi, S., and Lindsay, D. S. (1993). "Source Monitoring." *Psychological Bulletin,* 114, pp. 3–28.

47 Lillard, A. (2001) "Pretend Play as Twin Earth: A Social-Cognitive Analysis," *Developmental Review,* 21, pp. 495–531.

The idea that a banana is a phone in the imagination and a banana in reality has been called "double knowledge": McCune-Nicolich, L. (1981) "Toward Symbolic Functioning: Structure of Early Use of Early Pretend Games and Potential Parallels with Language," *Child Development,* 52, pp. 785-797.

Imagining hypothetical situations has been called "recreative imagination": Currie, G. and Ravenscroft, I. (2002) *Recreative Minds: Imagination in Philosophy and Psychology,* New York: Oxford University Press.

48 This is called the "single code" hypothesis.

49 Green, M. C., Brock, T. C., and Kaufman, G. F. (2004). "Understanding Media Enjoyment: The Role of Transportation into Narrative Worlds," *Communication Theory,* 14 (4), pp. 311–327.

50 My solution to this puzzle is described in my previous book, *Riveted.*

51 Nichols, S. (2004) "Imagining and Believing: The Promise of a Single Code." *Journal of Aesthetics and Art Criticism,* 62 (2), pp. 129–39.

52 Davies, J. and Bicknell, J. (2016). "Imagination and Belief: The Microtheories Model of Hypothetical Thinking." *Journal of Consciousness Studies,* 23(3–4), pp. 31–49.

53 Nichols, S. (Ed.). (2006). *The Architecture of the Imagination: New Essays on Pretence, Possibility, and Fiction.* Oxford: Oxford University Press. pp. 74.

54 Davies, J. and Bicknell, J. (2016). Imagination and belief: The microtheories model of hypothetical thinking. *Journal of Consciousness Studies,* 23(3-4), 31–49.

55 Davies, J. and Bicknell, J. (2016). "Imagination and Belief: The Microtheories Model of Hypothetical Thinking." *Journal of Consciousness Studies,* 23(3–4), 31–49.

56 This was discovered by the Cyc project, a many decades-long (and still ongoing) attempt to encode all human commonsense knowledge into a computer-readable database.

Lenat, D. B., and Guha, R. V. (1990) "Cyc: A Midterm Report," *AI Magazine,* 11 (3), pp. 32–59.

57 The illusions that your early visual system believes are charmingly called "aliefs" (in contrast with "b-liefs") by the philosopher Tamar Gendler.

Gendler, T. S. (2008) "Alief and belief," *The Journal of Philosophy,* 105 (10), pp.634–663.

58 Schacter, D. L., Addis, D. R., Hassabis, D., Martin, V. C., Spreng, R. N., and Szpunar, K. K. (2012). "The Future of Memory: Remembering, Imagining, and the Brain." *Neuron,* 76(21), pp. 677–694.

Schacter, D. L. (2012). "Adaptive Constructive Processes and the Future of Memory." *American Psychologist,* 67(8), p. 603.

59 Plants also have some other limited forms of memory. See Chamovitz, D. (2012). *What a Plant Knows: A Field Guide to the Senses.* New York: Scientific American Press. pp. 116–119.

60 Nairne, J. S., and Pandeirada, J. N. (2010). "Adaptive Memory: Nature's Criterion and the Functionalist Agenda." *The American Journal of Psychology,* 123(4), pp. 381–390.

3. IMAGINING THE FUTURE

1 Clayton, N. S., Bussey, T. J., and Dickinson, A. (2003). "Can Animals Recall the Past and Plan for the Future? *Nature Reviews Neuroscience,* 4, pp. 685–691.

2 D'Argembeau, A., Renaud, O., and Van der Linden, M. (2011). "Frequency, Characteristics and Functions of Future-Oriented Thoughts in Daily Life." *Applied Cognitive Psychology,* 25. pp. 96–103.

3 Being able to transport oneself into the past or future using imagination has been called "mental time travel," or "chronesthesia." Thinking about future events is sometimes called "prospection," as the opposite of "retrospection," which is thinking about the past.

Atance, C. M., and O'Neill, D. K. (2001). "Episodic Future Thinking." *Trends in Cognitive Science,* 5, pp. 533–539.

In further support of this, brain imaging studies show that imagining events in the past and future activate similar brain areas, where imagining something unrelated to personal experience in time (e.g., imagining Bill Clinton) does not: Szpunar, K. K., Watson, J. M., and McDermott, K. B. (2007). "Neural Substrates of Envisioning the Future." *Proceedings of the National Academy of Sciences of the United States of America,* 104, pp. 642–647. Further, individuals who lose their ability to recall past episodic memories also lose the ability to imagine their future: Klein, S. B., Loftus, J., and Kihlstrom, J. F. (2002). "Memory and Temporal Experience: The Effects of Episodic Memory Loss on an Amnesic Patient's Ability to Remember the Past and Imagine the Future. *Social Cognition,* 20, pp. 353–379.

4 Diseases: Schacter, D. L., Addis, D. R., Hassabis, D., Martin, C. C., Spreng, R. N., and Szpunar, K. K. (2012). "The Future of Memory: Remembering, Imagining, and the Brain." *Neuron,* 76(21), pp. 677–694. Though certain diseases, like Parkinson's and semantic dementia, can target one and not the other.

Older adults: Craik, F. I. M., and Salthouse, T. A. (Eds.). (2000). *Handbook of Aging and Cognition* (2nd ed.). Hillsdale, NJ: Erlbaum.

5 Schacter, D. L., Addis, D. R., Hassabis, D., Martin, C. C., Spreng, R. N., and Szpunar, K. K. (2012). "The Future of Memory: Remembering, Imagining, and the Brain. *Neuron,* 76(21), pp. 677–694.

Not all brain areas are the same for imagining the past and future, however. See pages 681–683 of this paper for a review.

6 Schkade, D. A., and Kahneman, D. (1998). Does living in California make people happy? A focusing illusion in judgments of life satisfaction. *Psychological Science,* 9, 340–346.

That study had almost two thousand people in it, and specifically found that life satisfaction scores were not affected by weather. However, Seth Stephens-Davidowitz found that during the winter, warm places like Honolulu had 40 percent fewer web searches for "depression" than cities like Chicago.

Stephens-Davidowitz, S. (2017). *Everybody Lies: Big Data, New Data, and What the Internet Can Tell Us about Who We Really Are.* New York: HarperCollins.

7 For each society, there seems to be a point where yearly income increases no longer are associated with happiness increases. For North America, this point has been estimated to be between $75,000 and $105,000. https://qz.com/1211957/how-much-money-do-people-need-to-be-happy/.

The idea that winning the lottery has no long-term effect on happiness has mixed results in the literature. https://www.businessinsider.com/lottery-winners-happiness-study-2018-8.

8 Paralyzed people often return to pre-paralysis levels of happiness after a few years, but people who are more agreeable adapt better to their new circumstances. "Agreeableness" is a core personality trait of human psychology.

Boyce. C. J., and Wood, A. M. (2011). Personality prior to disability determines adaptation: Agreeable individuals recover lost life satisfaction faster and more completely. *Psychological Science,* 22, pp. 1397–1402.

9 In general, the nature/nurture debate, when it comes to psychology, be it intelligence, happiness, or whatever, hovers around 50 percent one and 50 percent the other.

Lykken, D., and Tellegen, A. (1996). "Happiness Is a Stochastic Phenomenon." *Psychological science,* 7(3), pp. 186–189.

10 Beuhler, R., and McFarland, C. (2001). "Intensity Bias in Affective Forecasting: The Role of Temporal Focus." *Psychological Science,* 16, pp. 626–630.

I explore how this effect results in people having warped ideas about how the world works because of news media in my previous book, *Riveted.*

11 One way the world is clearly getting worse is climate change. For evidence of all the ways the world is or isn't improving, please read this excellent overview: Pinker, S. (2018). *Enlightenment Now: The Case for Reason, Science, Humanism, and Progress.* New York: Viking.

12 Read, D., and Lowenstein, G. F. (1995). "Diversification Bias: Explaining the Discrepancy in Variety Seeking between Combined and Separated Choices." *Journal of Experimental Psychology: Applied*, 1, pp. 34–49.

13 This concept is eloquently and hilariously described in one of my favorite books: Gilbert, D. (2006). *Stumbling on Happiness*. Toronto: Vintage Canada.

14 This is called the "planning bias." Nussbaum, S., Liberman, N., and Trope, Y. (2006). Predicting the near and distant future. *Journal of Experimental Psychology: General*, 135, pp. 152–161.

15 D'Argembeau, A., and Van der Linden, M. (2004). "Phenomenal Characteristics Associated with Projecting Oneself Back into the Past and Forward into the Future: Influence of Valence and Temporal Distance." *Consciousness and Cognition*, 13, pp. 844–858.

16 Underestimating future task completion times: Lam, K. C., Beuhler, R., McFarland, C., Ross, M., and Cheung, I. (2005). "Cultural Differences in Affective Forecasting: The Role of Focalism." *Personality and Social Psychology Bulletin*, 31, pp. 1296–1309.

Future events are imagined prototypically: Van Boven, L., Kane, J., and McGraw, A. P. (2009). "Temporally Asymmetric Constraints on Mental Simulation: Retrospection Is More Constrained than Prospection." In K. D. Markman, W. M. P. Klein, and J. A. Suhr (Eds.). *Handbook of Imagination and Mental Simulation*. New York: Taylor & Francis Group. pp. 131–147.

Anderson, R. J., and Dewhurst, S. A. (2009). "Remembering the Past and Imagining the Future: Differences in Event Specificity of Spontaneously Generated Thought." *Memory* 17, pp. 367–373.

17 Bavelas, J. B. (1973). "Effects of Temporal Context of Information." *Psychological Reports*, 32(3), pp. 695–698.

18 Pronin, E., and Ross, L. (2006). "Temporal Differences in Trait Self-Ascription: When the Self Is Seen as an Other. *Journal of Personality and Social Psychology*, 39, pp. 563–576.

Hershfield, H. E., and Bartels, D.M. (2018). "The Future Self." Chapter 5 of G. Oettingen, A. T. Sevincer, and P. M. Gollwitzer (Eds.). *Thinking About the Future*. New York: The Guilford Press.

19 Chen, M. K. (2013). "The Effect of Language on Economic Behavior: Evidence from Savings Rates, Health Behaviors, and Retirement Assets." *American Economic Review*, 103(2), pp. 690–731.

20 That part of the brain is the rostral cingulate cortex. Ross, V. (2010). "When I'm 64." *Scientific American Mind*, 21, p. 12.

21 Liberman, N., and Trope, Y. (1998). "The Role of Feasibility and Desirability Considerations in Near and Distant Future Decisions: A Test of Temporal Construal Theory." *Journal of Personality and Social Psychology*, 75, pp. 5–18.

22 Stephens-Davidowitz, S. (2017). *Everybody Lies: Big Data, New Data, and What the Internet Can Tell Us about Who We Really Are*. New York: HarperCollins.

23 Shu, S. B., and Gneezy, A. (2010). Procrastination of Enjoyable Experiences. *Journal of Marketing Research*, 47(5), pp. 933–934.

24 Loewenstein, G., Brennan, T., and Volpp, K. G. (2007). "Asymmetric Paternalism to Improve Health Behaviors." *Journal of the American Medical Association*, 298(20), pp. 2415–2417.

25 Money: Steinberg, L., Graham, S., O'Brien, L., Woolard, J., Cauffman, E., and Banich, M. (2009). "Age Differences in Future Orientation and Delay Discounting." *Child Development*, 80(1), pp. 28–44.

Environment: Strathman, A., Gleicher, F., Boninger, D. S., and Scott, E. C. (1994). "The Consideration of Future Consequences: Weighing Immediate and Distance Outcomes of Behavior." *Journal of Personality and Social Psychology*, 66(4), pp. 742–752.

26 Looking at old pictures of yourself makes you invest more in retirement: Ersner-Hershfield, H., Wimmer, G. E., and Knutson, B. (2009). "Saving for the Future Self: Neural Measures of the Future Self-Continuity Predict Temporal Discounting." *Social Cognitive and Affective Neuroscience*, 4, pp. 85–92.

Hershfield, H. E., Goldstein, D. G., Sharpe, W. F., Fox, J., Yeykelis, L., Carstensen, L. L., and Bailenson, J. N. (2011). "Increasing Saving Behaviour through Age-Progressed Renderings of the Future Self." *Journal of Marketing Research*, XLVIII, pp. s23–237.

27 Hershfield, H. E., Goldstein, D. G., Sharpe, W. F., Fox, J., Yeykelis, L., Carstensen, L. L., and Bailenson, J. N. (2011). "Increasing Saving Behavior through Age-Progressed Renderings of the Future Self. *Journal of Marketing Research*, 48(SPL), pp. S23–S37.

Blouin-Hudon, E., and Pychyl, T. A. (2017). "A Mental Imagery Intervention to Increase Future Self-Continuity and Reduce Procrastination." *Applied Psychology: An International Review*, 66(2), pp. 326–352.

28 Bandura, A. (1997). "The Anatomy of Stages of Change." *American Journal of Health Promotion*, 12, pp. 8–10.

29 Emmons, R. A., and Stern, R. (2013). "Gratitude as a Psychotherapeutic Intervention." *Journal of Clinical Psychology*, 69(8), pp. 846–855.

30 Taylor, S. E., Pham, L. B., Rivkin, I. D., and Armor, D. A. (1998). "Harnessing the Imagination: Mental Simulation, Self-Regulation, and Coping." *American Psychologist*, 53(4), pp. 429–439.

Programs designed to help people quit smoking often have a "public commitment" aspect, in which the smoker is encouraged to tell people that they are quitting smoking. In the limited literature search I did, however, although whole programs are often effective at reducing smoking, I have not found any study that isolated that variable.

31 Gollwitzer, P. M., Sheeran, P., Michalski, V., and Seifert, A. E. (2009). "When Intentions Go Public: Does Social Reality Widen the Intention-Behavior Gap?" *Psychological science*, 20(5), pp. 612–618.

32 Taylor, S. E., Pham, L. B., Rivkin, I. D., and Armor, D. A. (1998). "Harnessing the Imagination: Mental Simulation, Self-Regulation, and Coping." *American Psychologist*, 53(4), pp. 429–439.

Hershfield, H. E., and Bartels, D. M. (2018). "The Future Self." Chapter 5 of G. Oettingen, A. T. Sevincer, and P. M. Gollwitzer (Eds.). *Thinking about the Future*. New York: The Guilford Press. p. 101.

33 Gerlach, K. D., Spreng, R. N., Madore, K. P., and Schacter, D. L. (2014). "Future Planning: Default Network Activity Couples with Frontoparietal Control Network and Reward-Processing Regions During Process and Outcome Simulations. *Social Cognitive and Affective Neuroscience*, nsu001, pp. 1942–1951.

34 Predicting how you will feel in the future is called "affective forecasting." Overestimating the length and intensity of future emotions is called "impact bias." Gilbert, D. T., Pinel, E. C., Wilson, T. D., Blumberg, S. J., and Wheatley, T. P. (1998). "Immune Neglect: A Source of Durability Bias in Affective Forecasting." *Journal of Personality and Social Psychology*, 75, pp. 617–638.

Your affective forecasting is generally positive: Newby-Clark, I. R., and Ross, M. (2003). "Conceiving the Past and Future." *Personality and Social Psychology Bulletin*, 20, 807–818.

35 Van Boven, L., and Ashworth, L. (2007). "Looking Forward, Looking Back: Anticipation Is More Evocative than Retrospection." *Journal of Experimental Psychology: General,* 136, pp. 289–300. Gilbert, D. (2006). *Stumbling on Happiness.* Toronto: Vintage Canada. pp. 166–7.

36 Caruso, E. M. (2010). "When the Future Feels Worse than the Past: A Temporal Inconsistency in Moral Judgment." *Journal of Experimental Psychology: General* 139, pp. 610–624.

37 Caruso, E. M. (2010). "When the Future Feels Worse than the Past: A Temporal Inconsistency in Moral Judgement." Journal of Experimental Psychology, 139(4). pp. 610–624.

38 The Against Malaria Foundation has been rated as one of the most effective charities in the world, in terms of how many lives are saved per dollar donated. Thompson, Derek. "The Most Efficient Way to Save a Life." *The Atlantic.* The Atlantic Monthly Group. Retrieved 18 March 2017.

39 Small, D. A., Loewenstein, G., and Slovic, P. (2007). "Sympathy and Callousness: The Impact of Deliberative Thought on Donations to Identifiable and Statistical victims. *Organizational Behavior and Human Decision Processes,* 102, 143–153.

40 Dunn, E. W., and Ashton-James, C. (2008). "On Emotional Innumeracy: Predicted and Actual Affective Responses to Grand-Scale Tragedies." *Journal of Experimental Social Psychology,* 44(3), pp. 692–698, Study 3.

41 Singer, P. (2015). "The Logic of Effective Altruism." *Boston Review.* July 6, retrieved November 2018 from http://bostonreview.net/forum/peter-singer-logic-effective-altruism.

4. IMAGINATION, FEELINGS, AND MORALITY

1 Libby, L. K., and Eibach, R. P. (2009). "Seeing the Links among the Personal Past, Present, and Future: How Imagery Perspective in Mental Simulation Functions in Defining the Temporally Extended Self." In K. D. Markman, W. M. P. Klein, and J. A. Suhr (Eds.). *Handbook of Imagination and Mental Simulation.* New York: Taylor & Francis Group. pp. 359–372.

2 The ability to understand what is going on in someone else's head is known as theory of mind, using folk psychology, empathy, accurate empathy, cognitive empathy, social insight, and cognitive role-taking. Myers, M. W., and Hodges, S. D. (2009). "Making It up and Making Do: Simulation, Imagination, and Empathic Accuracy." In K. D. Markman, W. M. P. Klein, and J. A. Suhr (Eds.). *Handbook of Imagination and Mental Simulation.* New York: Taylor & Francis Group. pp. 281–293.

3 Epley, N., and Caruso, E. M. (2009). "Perspective Taking: Misstepping into Others' Shoes." In K. D. Markman, W. M. P. Klein, and J. A. Suhr (Eds.). *Handbook of Imagination and Mental Simulation.* New York: Taylor & Francis Group. pp. 295–309.

4 See also Bloom, P. (2016). *Against Empathy: The Case for Rational Compassion.* New York: Ecco.

5 Nichols, S., and Knobe, J. (2007). "Moral Responsibility and Determinism: The Cognitive Science of Folk Intuitions." *Nous,* 41, pp. 663–685.

6 Amit, E., and Greene, J. D. (2012). "You See, the Ends Don't Justify the Means: Visual Imagery and Moral Judgment." *Psychological Science,* 23(8), pp. 861–868.

7 For an entertaining podcast episode on what we don't understand about trolley problems, see the *Philosophy Bites* episode called "David Edmonds on Trolley Problems." http://philosophybites.com/2013/09/david-edmonds-on-trolley-problem.html.

8 There is evidence to suggest that imagery is strongly related to emotions. Mental imagery is associated with anxiety: Holmes, E. A., and Mathews, A. (2005). "Mental Imagery and Emotion: A Special Relationship?" *Emotion,* 5(4), p. 489.

And looking at pictures is more emotional than words: Holmes, E. A., Mathews, A., Mackintosh, B., and Dalgleish, T. (2008). "The Causal Effect of Mental Imagery on Emotion Assessed Using Picture-Word Cues." *Emotion*, 8(3), 395.

And people who are better imagers tend to be more emotional: Miller, G. A., Levin, D. N., Kozak, M. J., Cook III, E. W., McLean Jr, A., and Lang, P. J. (1987). "Individual Differences in Imagery and the Psychophysiology of Emotion." *Cognition and Emotion*, 1(4), 367–390.

9 Gaesser, B., and Schacter, D. L. (2014). "Episodic Simulation and Episodic Memory Can Increase Intentions to Help Others. *Proceedings of the National Academy of Sciences*, 111(12), 4415–4420.

10 Tasimi, A., and Young, L. (2016). "Memories of Good Deeds Past: The Reinforcing Power of Prosocial Behavior in Children." *Journal of Experimental Child Psychology*, 147, 159–166.

11 I talked about this in detail in the chapter on thinking about the future.

Taylor, S. E., Pham, L. B., Rivkin, I. D., and Armor, D. A. (1998). "Harnessing the Imagination: Mental Simulation, Self-Regulation, and Coping. *American Psychologist*, 53(4), 429.

12 The moral credential effect is also known as self-licensing, moral licensing, and the licensing effect.

Sachdeva, S., Iliev, R., and Medin, D. L. (2009). "Sinning Saints and Saintly Sinners: The Paradox of Moral Self-Regulation." *Psychological Science*, 20, 523–528.

Though there have been replication issues: Blanken, I., van de Ven, N., Zeelenberg, M., and Meijers, M. H. C. (2014). "Three Attempts to Replicate the Moral Licensing Effect." *Social Psychology*, 45(3), pp. 232–238.

That said, a meta-analysis (a study of many studies) found that the moral licensing effect was, across studies, a real but small- to medium-sized effect: Blanken, I., van de Ven, N., and Zeelenberg, M. (2015). "A Meta-Analytic Review of Moral Licensing." *Personality and Social Psychology Bulletin*, 41(4), pp. 540–558.

See also the Wikipedia page for "self-licensing" for a brief overview.

13 Conway, P., and Peetz, J. (2012). "When Does Feeling Moral Actually Make You a Better Person? Conceptual Abstraction Moderates whether Past Moral Deeds Motivate Consistency or Compensatory Behavior." *Personality and Social Psychology Bulletin*, 38(7), pp. 907–919.

14 There is counterevidence to this, though. Eugene Epley (experiment 2) found that mental simulation and imagery made people more generous in a dictator game (where they decided how much money they versus another person would get).

Caruso, E. M., and Gino, F. (2011). "Blind Ethics: Closing One's Eyes Polarizes Moral Judgments and Discourages Dishonest Behavior." *Cognition*, 118(2), pp. 280–285.

15 Cornelissen, G., Bashshur, M. R., Rode, J., and Le Menestrel, M. (2013). "Rules or Consequences? The Role of Ethical Mind-Sets in Moral Dynamics." *Psychological Science*, 24(4), pp. 482–488.

16 Blair, I. V., Ma, J. E., and Lenton, A. P. (2001). "Imagining Stereotypes Away: The Moderation of Implicit stereotypes through mental imagery. *Journal of Personality and Social Psychology*, 81, pp. 828–841.

17 Lecci, L., Okun, M. A., and Karoly, P. (1994). "Life Regrets and Current Goals as Predictors of Psychological Adjustment." *Journal of Personality and Social Psychology*, 66, 731–741.

18 Oettingen, G., Mayer, D., and Portnow, S. (2016). "Pleasure Now, Pain Later: Positive Fantasies about the Future Predict Symptoms of Depression." *Psychological Science*, 17(3), pp. 345–353.

19 If you think about how your life could be better, it's called an "upward counterfactual," and if you think about how your life could be worse, it's called a "downward counterfactual."
Markman, K. D., Gavanski, I., Sherman, S. J., and McMullen, M. N. (1993). "The Mental Simulation of Better and Worse Possible Worlds." *Journal of Experimental Social Psychology*, 29, pp. 87–109.

20 Markman, K. D., Karadogan, F., Lindberg, M. J., and Zell, E. (2009). "Counterfactual Thinking: Function and Dysfunction." In K. D. Markman, W. M. P. Klein, and J. A. Suhr (Eds.). *Handbook of Imagination and Mental Simulation.* New York: Taylor & Francis Group. pp. 175–193.

21 Emmons, R. A., and Stern, R. (2013). "Gratitude as a Psychotherapeutic Intervention." *Journal of Clinical Psychology*, 69(8), 846–855.

22 Byrne, R. M. J., and Girotto, V. (2009). "Cognitive processes in counterfactual thinking." In K. D. Markman, W. M. P. Klein, and J. A. Suhr (Eds.). *Handbook of Imagination and Mental Simulation.* New York: Taylor & Francis Group. pp. 151–160.

23 Wong, E. M., Galinsky, A. D., and Kray, L. J. (2009). "The Counterfactual Mind-Set: A Decade of Research." In K. D. Markman, W. M. P. Klein, and J. A. Suhr (Eds.). *Handbook of Imagination and Mental Simulation.* New York: Taylor & Francis Group. pp. 161–174.

24 Green, M. C., and Brock, T. C. (2000). "The Role of Transportation in the Persuasiveness of Public Narratives." *Journal of Personality and Social Psychology*, 79, pp. 701–721.

25 His name is approximately pronounced "mee-HIGH cheeck-SENT-me-high."
Csikszentmihalyi, M. (1990). *Flow: The Psychology of Optimal Experience.* New York: Harper and Row.

26 Green, M. C., Brock, T. C., and Kaufman, G. F. (2004). "Understanding Media Enjoyment: The Role of Transportation into Narrative Worlds." *Communication Theory*, 14(4), pp. 311–327.

27 Moskalenko, S., and Heine, S. J. (2003). "Watching Your Troubles Away: Television Viewing as a Stimulus for Subjective Self-Awareness." *Personality and Social Psychology Bulletin*, 29(1), pp. 76–85.

28 Green, M. C. (2004). "Transportation into Narrative Worlds: The Role of Prior Knowledge and Perceived Realism." *Discourse Processes*, 38(2), pp. 247–266.

29 Marsh, E. F., and Fazio, L. K. (2006). "Learning Errors from Fiction: Difficulties in Reducing Reliance on Fictional Stories." *Memory & Cognition*, 34(5), pp. 1140–1149.

30 Pinker, S. (2011). *The Better Angels of our Nature: Why Violence Had Declined.* New York: Viking.

31 Mar, R. A., Oatley, K., Hirsh, J., dela Paz, J., and Peterson, J. B. (2006). "Bookworms Versus Nerds: Exposure to Fiction Versus Nonfiction, Divergent Associations with Social Ability, and the Simulation of Fictional Social Worlds." *Journal of Research in Personality*, 40(5), pp. 694–712.

32 Katz, P. A., and Zalk, S. R. (1978). "Modification of Children's Racial Attitudes." *Developmental Psychology*, 14(5), pp. 447.

5. HALLUCINATION

1 Peyroux, E., and Franck, N. (2013). "An Epistemological Approach: History of Concepts and Ideas about Hallucinations in Classical Psychiatry." In R. Jardri, A. Cachia, P. Thomas, and D. Pins (Eds.). *The Neuroscience of Hallucinations.* New York: Springer. pp. 3–20.

2 Peyroux, E., and Franck, N. (2013). "An Epistemological Approach: History of Concepts and Ideas about Hallucinations in Classical Psychiatry." In R. Jardri, A. Cachia, P. Thomas, and D. Pins (Eds.). *The Neuroscience of Hallucinations.* New York: Springer. pp. 3–20, 8.

3 Larøi, F., Luhrmann, T. M., Bell, V., Christian Jr, W. A., Deshpande, S., Fernyhough, C., and Woods, A, et. al. (2014). "Culture and Hallucinations: Overview and Future Directions." *Schizophrenia Bulletin*, 40 (Suppl. No. 4), pp. S213–S220.

4 Rees, D. W. (1971). "The Hallucinations of Widowhood." *British Medical Journal*, 4, pp. 37–41.

5 Sacks, O. (2012). *Hallucinations*. New York: Alfred A. Knopf. pp. 253–254.

6 Blom, J. D. (2013). "Hallucinations and Other Sensory Deceptions in Psychiatric Disorders." In R. Jardri, A. Cachia, P. Thomas, and D. Pins (Eds.). *The Neuroscience of Hallucinations*. New York: Springer. pp. 42–57, 45.

7 Weiss, A. P., and Heckers, S. (1999). "Neuroimaging of Hallucinations: A Review of the Literature." *Psychiatry Research: Neuroimaging*, 92(2–3), pp. 61–74.

This is also known as "reality discrimination:" Bentall, R. P. (1990). "The Illusion of Reality: A Review and Integration of Psychological Research on Hallucinations." *Psychological Bulletin*, 107(1), p. 82.

8 Boissier de Sauvages believed this in 1768, suggesting that hallucinators mistook their perceptions for imagery and vice versa. Peyroux, E., and Franck, N. (2013). "An Epistemological Approach: History of Concepts and Ideas about Hallucinations in Classical Psychiatry." In R. Jardri, A. Cachia, P. Thomas, and D. Pins (Eds.). *The Neuroscience of Hallucinations*. New York: Springer. pp. 3–20, 7.

9 Barber, T. X., and Calverey, D. S. (1964). "An Experimental Study of Hypnotic (Auditory and Visual) Hallucinations." *Journal of Abnormal and Social Psychology*, 68, pp. 13–20.

10 Sacks, O. (2012). *Hallucinations*. New York: Alfred A. Knopf.

11 Aleman, A., and Vercammen, A. (2013). "The 'Bottom-Up' and "Top-Down' Components of the Hallucinatory Phenomenon." In R. Jardri, A. Cachia, P. Thomas, and D. Pins (Eds.). *The Neuroscience of Hallucinations*. New York: Springer. pp. 107–121, 119.

12 Sacks, O. (2012). *Hallucinations*. New York: Alfred A. Knopf. pp. 23–24.

13 Sacks, O. (2012). *Hallucinations*. New York: Alfred A. Knopf. pp. 41–42.

14 This is sometimes known as the "Third Man Factor." Shackleton, E. H. (2004). *South: The Story of Shackleton's 1914–1917 Expedition*. Mobile Bounty Pty Ltd.

15 Sacks, O. (2012). *Hallucinations*. New York: Alfred A. Knopf. pp. 81–83.

16 Sacks, O. (2012). *Hallucinations*. New York: Alfred A. Knopf. pp. 149–150.

17 Burton, R.A. (2017). "When Neurology becomes Theology." *Nautilus*, May/June, pp. 88–94.

18 Sacks, O. (2012). *Hallucinations*. New York: Alfred A. Knopf. p. 258.

19 Sapolsky, R. (2016). To Understand Facebook, Study Capgras Syndrome." *Nautilus*, Nov/Dec, pp. 52–59.

20 Around 20 percent of Alzheimer's patients experience hallucinations. Fénelon, G. (2013). "Hallucinations Associated with Neurological Disorders and Sensory loss. In R. Jardri, A. Cachia, P. Thomas, and D. Pins (Eds.). *The Neuroscience of Hallucinations*. New York: Springer. pp. 59–83, 62.

Sacks, O. (2012). *Hallucinations*. New York: Alfred A. Knopf. p. 87.

Recent neuroscience work has associated Capgras syndrome with problems with connectivity between the areas of the brain involved with familiarity (the left retrospinial cortex) and those associated with expectation violation (the right frontal cortex).

Darby, R. R., Laganiere, S., Pascual-Leone, A., Prasad, S., and Fox, M. D. (2017). "Finding the Imposter: Brain Connectivity of Lesions Causing Delusional Misidentifications." *Brain*, 140, pp. 497–507.

21 Sixty to eighty percent of patients with schizophrenia experience auditory verbal hallucinations. Less than 10 percent experience visual hallucinations. Peyroux, E., and Franck, N. (2013). "An Epistemological Approach: History of Concepts and Ideas about Hallucinations in Classical Psychiatry." In R. Jardri, A. Cachia, P. Thomas, and D. Pins (Eds.). *The Neuroscience of Hallucinations.* New York: Springer. pp. 3–20.

Sacks, O. (2012). *Hallucinations.* New York: Alfred A. Knopf. pp. 57–58.

Fénelon, G. (2013). "Hallucinations Associated with Neurological Disorders and Sensory Loss." In R. Jardri, A. Cachia, P. Thomas, and D. Pins (Eds.). *The Neuroscience of Hallucinations.* New York: Springer. pp. 59–83, 61.

22 Hill, K., and Linden, D. E. J. (2013). Hallucinatory Experiences in Non-clinical Populations. In R. Jardri, A. Cachia, P. Thomas, and D. Pins (Eds.). *The Neuroscience of Hallucinations.* New York: Springer. pp. 21–41.

23 Forty-six to seventy-three percent of patients with dementia (with Lewy bodies) experience hallucinations. Fénelon, G. (2013). "Hallucinations Associated with Neurological Disorders and Sensory Loss." In R. Jardri, A. Cachia, P. Thomas, and D. Pins (Eds.). *The Neuroscience of Hallucinations.* New York: Springer. pp. 59–83, 62.

Hill, K., and Linden, D. E. J. (2013). "Hallucinatory Experiences in Non-Clinical Populations." In R. Jardri, A. Cachia, P. Thomas, and D. Pins (Eds.). *The Neuroscience of Hallucinations.* New York: Springer. pp. 21–41, 22.

24 Hill, K., and Linden, D. E. J. (2013). "Hallucinatory Experiences in Clinical Populations." In R. Jardri, A. Cachia, P. Thomas, and D. Pins (Eds.). *The Neuroscience of Hallucinations.* New York: Springer. pp. 21–41, 22.

25 The phenomena of hearing voices that aren't real is known as "auditory verbal hallucination." It happens to healthy people but is also a symptom of several mental illnesses. Auditory verbal hallucinations are symptoms of schizophrenia, depressive disorder, postpartum psychosis, bipolar disorder, post-traumatic stress disorder, delirium and delirium tremens, alcoholic hallucinosis, dementia, many types of substance-abuse disorders, conversion, dissociative disorder, and borderline personality disorder. Blom, J. D. (2013). Hallucinations and other sensory deceptions in psychiatric disorders. In R. Jardri, A. Cachia, P. Thomas, and D. Pins (Eds.). *The Neuroscience of Hallucinations.* New York: Springer. pp. 42–57, 46.

It's clear that during verbal hallucinations something is not working in the brain, but we don't yet know what that is. But there are theories, and each one has some evidence for it. Some think that it's a disorder in speech planning, others a problem in speech perception, or recognizing that the inner speech is yours, or recognizing what is real versus make-believe, or recognizing that you have taken action.

Peyroux, E., and Franck, N. (2013). "An Epistemological Approach: History of Concepts and Ideas about Hallucinations in Classical Psychiatry." In R. Jardri, A. Cachia, P. Thomas, and D. Pins (Eds.). *The Neuroscience of Hallucinations.* New York: Springer. pp. 3–20, 16–17.

There are a surprising number of ways you can have voices in your head. One is called *gedankenlautwerden*, which must be rare because we don't even have an English word for it. It literally means "thoughts becoming loud" in German, and is the experience of hearing words before you think them. Indirect *gedankenlautwerden* is when you attribute your own thoughts to the content of other people's voices.

Blom, J. D. (2013). "Hallucinations and Other Sensory Deceptions in Psychiatric Disorders." In R. Jardri, A. Cachia, P. Thomas, and D. Pins (Eds.). *The Neuroscience of Hallucinations.* New York: Springer. pp. 42–57, 46.

26 Peyroux, E., and Franck, N. (2013). "An epistemological approach: History of concepts and ideas about hallucinations in classical psychiatry. In R. Jardri, A. Cachia, P. Thomas, and D. Pins (Eds.). *The Neuroscience of Hallucinations.* New York: Springer. pp. 3–20, 15–16.

27 I have an episode about voices in the head on my podcast, *Minding the Brain.*

Peyroux, E., and Franck, N. (2013). An epistemological approach: History of concepts and ideas about hallucinations in classical psychiatry. In R. Jardri, A. Cachia, P. Thomas, and D. Pins (Eds.). *The neuroscience of hallucinations.* New York, NY: Springer. pp. 3-20, 16.

28 Hill, K., and Linden, D.E.J. (2013). Hallucinatory experiences in non-clinical populations. In R. Jardri, A. Cachia, P. Thomas, and D. Pins (Eds.). *The neuroscience of hallucinations.* New York: Springer. pp. 21-41, 22, 28.

29 Jenner, J. A. (2013). "Beyond Monotherapy: The HIT Story." In R. Jardri, A. Cachia, P. Thomas, and D. Pins (Eds.). *The Neuroscience of Hallucinations.* New York: Springer. pp.447–470, 449.

30 Lowe, N. G., Rapagnani, M. P., Mattei, C., and Stahl, S. M. (2013). "The Psychopharmacology of Hallucinations: Ironic Insights and Mechanisms of Action." In R. Jardri, A. Cachia, P. Thomas, and D. Pins (Eds.). *The Neuroscience of Hallucinations.* New York: Springer. pp. 471–492, 475.

31 Watkins, M. (2000). *Invisible Guests: The Development of Imaginal Dialogues.* Woodstock, CT: Spring Press. p. 115.

32 Glausiusz, J. (2014). "Living in an Imaginary World." *Scientific American Mind*, winter, special issue on creativity. pp. 70–77.

33 Sacks, O. (2012). *Hallucinations.* New York: Alfred A. Knopf.

34 Peyroux, E., and Franck, N. (2013). "An Epistemological Approach: History of Concepts and Ideas about Hallucinations in Classical Psychiatry." In R. Jardri, A. Cachia, P. Thomas, and D. Pins (Eds.). *The Neuroscience of Hallucinations.* New York,: Springer. pp. 3–20, 13.

35 Fénelon, G. (2013). "Hallucinations Associated with Neurological Disorders and Sensory Loss." In R. Jardri, A. Cachia, P. Thomas, and D. Pins (Eds.). *The Neuroscience of Hallucinations.* New York: Springer. pp. 59–83, 67.

36 Braithwaite, J. J., Samson, D., Apperly, I., Broglia, E., and Hulleman, J. (2011). "Cognitive Correlates of the Spontaneous Out-of-Body Experience (OBE) in the Psychologically Normal Population: Evidence for an Increased Role of Temporal-Lobe Instability, Body-Distortion Processing, and Impairments in Own-Body Transformations." *Cortex*, 47(7), 839–853.

37 Sacks, O. (2012). *Hallucinations.* New York: Alfred A. Knopf. pp. 90–98.

38 See a relevant case study in Sacks, O. (2012). *Hallucinations.* New York: Alfred A. Knopf. pp. 101.

39 Sacks, O. (2012). *Hallucinations.* New York: Alfred A. Knopf. pp. 106–107.

40 Christopher Timmermann, personal communication, 2018.

41 About 13 percent of people with epilepsy get hallucinations. Fénelon, G. (2013). "Hallucinations Associated with Neurological Disorders and Sensory Loss. In R. Jardri, A. Cachia, P. Thomas, and D. Pins (Eds.). *The neuroscience of Hallucinations.* New York: Springer. pp.59–83, 62.

Sacks, O. (2012). *Hallucinations.* New York: Alfred A. Knopf. p. 133.

42 Sacks, O. (2012). *Hallucinations.* New York: Alfred A. Knopf. p. 138.

43 Sacks, O. (2012). *Hallucinations.* New York: Alfred A. Knopf. p. 11.

44 Sacks, O. (2012). *Hallucinations.* New York: Alfred A. Knopf. pp. 12–13.

45 Sacks, O. (2012). *Hallucinations.* New York: Alfred A. Knopf. pp. 25, 27.

46 Burke, W. (2002). "The Neural Basis of Charlès Bonnet Hallucinations: A Hypothesis." *Journal of Neurology, Neurosurgery, and Psychiatry,* pp. 73, 535–541.

47 Under conditions of sensory deprivation, the hallucinations get more vivid over time: Sacks, O. (2012). *Hallucinations.* New York: Alfred A. Knopf. pp. 22, 26, 36–37.

48 Aleman, A., and Vercammen, A. (2013). "The 'Bottom-Up' and 'Top-Down' Components of the Hallucinatory Phenomenon." In R. Jardri, A. Cachia, P. Thomas, and D. Pins (Eds.). *The Neuroscience of Hallucinations.* New York: Springer. pp. 107–12, 114.

49 Sacks, O. (2012). *Hallucinations.* New York: Alfred A. Knopf. pp. 27–28 .

50 This is known as a "homunculus" in cognitive science.
Dennett, D. (1992). *Consciousness Explained.* Cambridge, MA: Back Bay Books.

51 Sacks, O. (2012). *Hallucinations.* New York: Alfred A. Knopf. p. 73.

52 Van Swam, C., Dierks, T., and Hubl, D. (2013). "Electrophysiological Exploration of Hallucinations." In R. Jardri, A. Cachia, P. Thomas, and D. Pins (Eds.). *The Neuroscience of Hallucinations.* New York: Springer. pp. 317–342, 333.

53 Sacks, O. (2012). *Hallucinations.* New York: Alfred A. Knopf. pp. 40–41.

54 About 20 percent of people with migraines get hallucinations, usually of the elementary visual kind, and sometimes somatosensory. Complex hallucinations (people or objects) are almost never reported with migraines. In migraine auras, a wave of neuronal activity triggers cells that detect orientation, resulting in the hallucinated perception of zigzag lines. Fénelon, G. (2013). "Hallucinations Associated with Neurological Disorders and Sensory Loss." In R. Jardri, A. Cachia, P. Thomas, and D. Pins (Eds.). *The Neuroscience of Hallucinations.* New York: Springer. pp. 59–83, 62, 65.
Sacks, O. (2012). *Hallucinations.* New York: Alfred A. Knopf. pp. 130–131.

55 About 22–38 percent of Parkinson's patients also experience hallucinations. Fénelon, G. (2013). "Hallucinations Associated with Neurological Disorders and Sensory Loss." In R. Jardri, A. Cachia, P. Thomas, and D. Pins (Eds.). *The Neuroscience of Hallucinations.* New York: Springer. pp. 59–83, 62.
Sacks, O. (2012). *Hallucinations.* New York: Alfred A. Knopf. p. 229.

56 Sacks, O. (2012). *Hallucinations.* New York: Alfred A. Knopf. p. 168.

57 Burke, W. (2002). The Neural Basis of Charles Bonnet Hallucinations: A Hypothesis. *Journal of Neurology and Neurosurgical Psychology,* 73. pp. 535–541.

58 Sacks, O. (2012). *Hallucinations.* New York: Alfred A. Knopf. pp. 237–238.

59 Zadra, A. L., Nielsen, T. A., and Donderi, D. C. (1998). "Prevalence of Auditory, Olfactory, and Gustatory Experiences in Home Dreams. *Perceptual and Motor Skills* 87, pp. 819–826.

60 Olfactory hallucinations are called "phantosmia." Sacks, O. (2012). *Hallucinations.* New York: Alfred A. Knopf. pp. 52.

61 Sacks, O. (2012). *Hallucinations.* New York: Alfred A. Knopf. pp. 47–48.

62 Hallucinations regarding body image are called "coenesthesiopathy" or "paracoenesthesiopathy," and 18 percent of people with schizophrenia have it. Hypercoenesthesiopathy is when you sense your body is gigantic, hypocoenesthesiopathy is when you feel it's very small, and acoenesthesiopathy is when you hallucinate that you have no body at all. Blom, J. D. (2013). "Hallucinations and Other Sensory Deceptions in Psychiatric Disorders." In R. Jardri, A. Cachia, P. Thomas, and D. Pins (Eds.). *The Neuroscience of Hallucinations.* New York: Springer. pp. 42–57, 55.

63 This is Anton-Babinski syndrome. Sacks, O. (2012). *Hallucinations.* New York: Alfred A. Knopf. p. 177.

64 Fénelon, G. (2013). "Hallucinations Associated with Neurological Disorders and Sensory Loss." In R. Jardri, A. Cachia, P. Thomas, and D. Pins (Eds.). *The Neuroscience of Hallucinations.* New York: Springer. pp. 59–83 69.

65 Peyroux, E., and Franck, N. (2013). "An Epistemological Approach: History of Concepts and Ideas about Hallucinations in Classical Psychiatry." In R. Jardri, A. Cachia, P. Thomas, and D. Pins (Eds.). *The Neuroscience of Hallucinations.* New York: Springer. pp. 3–20.

66 Jardri, R., Thomas, P., Pins, D., and Cachia, A. (2013). "Key Issues for Future Research in the Neuroscience of Hallucinations." In R. Jardri, A. Cachia, P. Thomas, and D. Pins (Eds.). *The Neuroscience of Hallucinations.* New York: Springer. pp. 549–552, 550.

67 Luhrmann, T. M. (2012). *When God Talks Back: Understanding the American Evangelical Relationship with God.* New York: Knopf.

68 Larøi, F., Luhrmann, T. M., Bell, V., Christian Jr., W. A., Deshpande, S., Fernyhough, C., and Woods, A, et al. (2014). "Culture and Hallucinations: Overview and Future Directions." *Schizophrenia Bulletin*, 40 (Suppl. No. 4), S213–S220.

69 Aleman, A., and Vercammen, A. (2013). "The 'Bottom-Up' and 'Top-Down' Components of the Hallucinatory Phenomenon." In R. Jardri, A. Cachia, P. Thomas, and D. Pins (Eds.). *The Neuroscience of Hallucinations.* New York: Springer. pp. 107–121, 117.

70 Larøi, F., Luhrmann, T. M., Bell, V., Christian Jr., W. A., Deshpande, S., Fernyhough, C., and Woods, A., et al. al. (2014). "Culture and Hallucinations: Overview and Future Directions." *Schizophrenia Bulletin*, 40 (Suppl. No. 4), pp. S213–S220.

71 Dein, S., and Littlewood, R. (2007). "The Voice of God." *Anthropology & Medicine*, 14(2), pp. 213–228.

Note what is happening here: the church members used their ideas of what God is like to determine which messages they hear are from God. If what they hear is not in accordance with what they believe God *would* say, the voice is dismissed as mere hallucination. This provides a feedback loop of "evidence" of what God is like that even God might not be able to break.

72 Larøi, F., Luhrmann, T. M., Bell, V., Christian Jr., W. A., Deshpande, S., Fernyhough, C., and Woods, A, et al. (2014). "Culture and Hallucinations: Overview and Future Directions." *Schizophrenia Bulletin*, 40 (Suppl. No. 4), pp. S213–S220.

73 Adler, S. R. (2011). *Sleep Paralysis: Night-Mares, Nocebos, and the Mind-Body Connection.* New Brunswick, NJ: Rutgers University Press.

I discuss the effects of sleep paralysis on belief in greater detail in my book *Riveted.*

74 Specifically, dementia with Lewy bodies. See Fénelon, G. (2013). "Hallucinations Associated with Neurological Disorders and Sensory Loss." In R. Jardri, A. Cachia, P. Thomas, and D. Pins (Eds.). *The Neuroscience of Hallucinations.* New York,: Springer. pp. 59–83, 73.

75 Gilles Fénelon, personal communication.

76 Fénelon, G. (2013). "Hallucinations Associated with Neurological Disorders and Sensory Loss." In R. Jardri, A. Cachia, P. Thomas, and D. Pins (Eds.). *The Neuroscience of Hallucinations.* New York: Springer. pp. 59–83, 63.

77 Bexton, W. H., Heron, W., and Scott, T. H. (1954). "Effects of Decreased Variation in the Sensory Environment. *Canadian Journal of Psychology,* 8(2), 70–76.

78 Fénelon, G. (2013). "Hallucinations Associated with Neurological Disorders and Sensory Loss." In R. Jardri, A. Cachia, P. Thomas, and D. Pins (Eds.). *The Neuroscience of Hallucinations.* New York: Springer. pp. 59–83, 63–64.

79 80 percent of AVH hearers report that symptoms are worse when they are alone. Rossell, S. (2013). "The Role of Memory Retrieval and Emotional Salience in the Emergence of

Auditory Hallucinations." In R. Jardri, A. Cachia, P. Thomas, and D. Pins (Eds.). *The Neuroscience of Hallucinations*. New York: Springer. pp. 137–151, 147.

80 Eveleth, R. (2014). "The Ancient, Peaceful Art of Self-Generated Hallucination." *Nautilus*, retrieved January 30, 2018, from http://nautil.us/blog/the-ancient-peaceful-art-of-self _generated-hallucination.

81 Jardri, R., and Denève, S. (2013). "Computational models of Hallucinations." In R. Jardri, A. Cachia, P. Thomas, and D. Pins (Eds.). *The Neuroscience of Hallucinations*. New York, NY: Springer. pp. 289–313, 302.

82 Jardri, R., and Denève, S. (2013). "Circular Inferences in Schizophrenia." *Brain, 136,* 3227–3241.

83 . . .activating brown, activating peanut butter, activating brown, activating peanut butter, activating brown, activating peanut butter, activating brown, activating peanut butter, activating brown, activating peanut butter, activating brown, activating peanut butter, activating brown, activating peanut butter, activating brown, activating peanut butter, activating brown, activating peanut butter.

84 Mobbs, D., Weiskopf, N., Lau, H. C., Featherstone, E., Dolan, R. J., and Frith, C. D. (2006). "The Kuleshov Effect: the Influence of Contextual Framing on Emotional Attributions. *Social cognitive and affective neuroscience*, 1(2), 95–106.

85 Corlett, P. R., Krystal, J. H., Taylor, J. R., and Fletcher, P. C. (2009). "Why Do Delusions Persist?" *Frontiers in Human Neuroscience*, 3, 12.

86 Denève, S., and Jardri, R. (2016). "Circular inference: Mistaken belief, misplaced trust. *Current Opinion in Behavioral Sciences*, 11, 40–48.

Jardri, R., Duverne, S., Litvinova, A. S., and Denève, S. (2017). Experimental evidence for circular inference in schizophrenia. Nature Communications, 8.

87 Rossell, S. (2013). The role of memory retrieval and emotional salience in the emergence of auditory hallucinations. In R. Jardri, A. Cachia, P. Thomas, and D. Pins (Eds.). The neuroscience of hallucinations. (pp. 137–151). New York, NY: Springer. Page 141.

88 Misattribution of an internally generated experience to an external source is known as "source monitoring," "reality discrimination," "source memory," and "self-monitoring." Rossell, S. (2013). "The Role of Memory Retrieval and Emotional Salience in the Emergence of Auditory Hallucinations." In R. Jardri, A. Cachia, P. Thomas, and D. Pins (Eds.). *The Neuroscience of Hallucinations.* New York: Springer. pp. 137–151, 142.

89 Rossell, S. (2013). "The Role of Memory Retrieval and Emotional Salience in the Emergence of Auditory Hallucinations." In R. Jardri, A. Cachia, P. Thomas, and D. Pins (Eds.). *The Neuroscience of Hallucinations*. New York: Springer. pp. 137–151, 1420–146.

90 This theory is called "dysfunction of the forward-model system." See Waters, F. (2013). "Time Perception and Discrimination in Individuals Suffering from Hallucinations." In R. Jardri, A. Cachia, P. Thomas, and D. Pins (Eds.). *The Neuroscience of Hallucinations*. (). New York: Springer. pp. 185–199, 191–192.

91 Blakemore, S.-J., Wolpert, D., and Frith, C. (1998). "Central Cancellation of Self-Produced Tickle Sensation." *Nature Neuroscience,* 1(7), pp. 635–640.

92 Panksepp, J., and Burgdorf, J. (2003). "'Laughing' Rats and the Evolutionary Antecedents of Human Joy?" *Physiology & Behavior*, 79, pp. 533–547.

93 Blakemore, S.-J., Wolpert, D., and Frith, C. (1998). "Central Cancellation of Self-Produced Tickle Sensation." *Nature Neuroscience,* 1(7), pp. 635–640.

94 Blakemore, S.-J., Smith, J., Steel, R., Johnstone, E., and Frith, C. (2000). "The Perception of Self-Produced Sensory Stimuli in Patients with Auditory Hallucinations and Passivity

Experiences: Evidence for a Breakdown in Self-Monitoring." *Psychological Medicine*, 30, pp. 1131–1139.

95 Kornmüller, A. E. (1931). "Eine experimentelle Anästhesie der äusseren Augenmuskeln am Menschen und ihre Auswirkungen." *J. Psychol. Neurol. (Lpz.)*, 41, pp. 354–366.
cited in: Brindley, G. S., and Merton, P. A. (1960). "The Absence of Position Sense in the Human Eye." *The Journal of Physiology*, 153(1), pp. 127–130.

96 Hill, K., and Linden, D. E. J. (2013). "Hallucinatory Experiences in Non-Clinical Populations." In R. Jardri, A. Cachia, P. Thomas, and D. Pins (Eds.). *The Neuroscience of Hallucinations*. New York: Springer. pp. 21–41, 34.

97 Van Swam, C., Dierks, T., and Hubl, D. (2013). "Electrophysiological Exploration of Hallucinations." In R. Jardri, A. Cachia, P. Thomas, and D. Pins (Eds.). *The Neuroscience of Hallucinations*. New York: Springer. pp. 317–342, 317.

98 The copy is called the "forward model," the "efference copy," or "corollary discharge." Van Swam, C., Dierks, T., and Hubl, D. (2013). "Electrophysiological Exploration of Hallucinations." In R. Jardri, A. Cachia, P. Thomas, and D. Pins (Eds.). *The Neuroscience of Hallucinations*. New York: Springer. pp. 317–342, 336.
The copy is called the "forward model," or "efference copy." The expectation in the sensory areas as a result of the copy arriving is the "corollary discharge."
Ford, J. M., and Hofman, R. E. (2013). "Functional Brain Imaging of Auditory Hallucinations: From Self-Monitoring Deficits to Co-opted Neural Resources." In R. Jardri, A. Cachia, P. Thomas, and D. Pins (Eds.). *The Neuroscience of Hallucinations*. New York: Springer. pp. 359–373, 362.

99 "Experiential hallucinations." Fénelon, G. (2013). "Hallucinations Associated with Neurological Disorders and Sensory Loss." In R. Jardri, A. Cachia, P. Thomas, and D. Pins (Eds.). *The Neuroscience of Hallucinations*. New York: Springer. pp. 59–83, 66.

100 Fénelon, G. (2013). "Hallucinations Associated with Neurological Disorders and Sensory Loss." In R. Jardri, A. Cachia, P. Thomas, and D. Pins (Eds.). *The Neuroscience of Hallucinations*. New York: Springer. pp. 59–83, 71–2.

101 Blom, J. D. (2013). "Hallucinations and Other Sensory Deceptions in Psychiatric Disorders." In R. Jardri, A. Cachia, P. Thomas, and D. Pins (Eds.) *The Neuroscience of Hallucinations*. New York: Springer. pp. 42–57, 47.

6. DREAMING

1 Hobson, J. A., Pace-Schott, E. F., and Stickgold, R. (2000). "Dreaming and the Brain: Toward a Cognitive Neuroscience of Conscious States." *Behavioral and Brain Sciences*, 23(06), pp. 793–842, 825.

2 Lynch, G., Colgin, L. L., and Palmer, L. (2000). "Spandrels of the night?" *Behavioral and Brain Sciences*, 23(06), pp. 966–7.

3 Sforza, E., Krieger, J., and Petiau, C. (1997). "REM Sleep Behavior Disorder: Clinical and Physiopathological Findings." *Sleep Medicine Reviews*, 1(1), pp. 57–69.

4 Hobson, J. A., Pace-Schott, E. F., and Stickgold, R. (2000). "Dreaming and the Brain: Toward a Cognitive Neuroscience of Conscious States." *Behavioral and Brain Sciences*, 23(06), pp. 793–842, 828–831.

5 Though roughly a quarter of REM-like dreams happen in NREM states: Hobson, J. A., Pace-Schott, E. F., and Stickgold, R. (2000). "Dreaming and the Brain: Toward a Cognitive Neuroscience of Conscious States." *Behavioral and Brain Sciences*, 23(06), pp. 793–842, 805, 844–5.

6 Sleepers awakened during REM report dreaming 70–95 percent of the time, as opposed to 5–10 percent in NREM states: Hobson, J. A., Pace-Schott, E. F., and Stickgold, R. (2000). "Dreaming and the Brain: Toward a Cognitive Neuroscience of Conscious States." *Behavioral and Brain Sciences*, 23(06), pp. 793–842, 843.

7 Hobson, J.A., Pace-Schott, E. F., and Stickgold, R. (2000). "Dream Science 2000: A Response to Commentaries on *Dreaming and the Brain*." *Behavioral and Brain Sciences*, 23(06), pp. 1019–1035.

8 Hobson, J. A., Pace-Schott, E. F., and Stickgold, R. (2000). "Dreaming and the Brain: Toward a Cognitive Neuroscience of Conscious States." *Behavioral and Brain Sciences*, 23(06), pp. 793–842, 802.

9 Hobson, J. A., Pace-Schott, E. F., and Stickgold, R. (2000). "Dreaming and the Brain: Toward a Cognitive Neuroscience of Conscious States." *Behavioral and Brain Sciences*, 23(06), pp. 793–842, 805.

10 Hobson, J. A., Pace-Schott, E. F., and Stickgold, R. (2000). "Dreaming and the Brain: Toward a Cognitive Neuroscience of Conscious States." *Behavioral and Brain Sciences*, 23(06), pp. 793–842, 802, 899.

11 Skaggs, W. E., an McNaughton, B. L. (1996). "Replay of Neuronal Firing Sequences in Rat Hippocampus during Sleep following Spatial Experience." *Science*, 271(5257), p. 1870.

12 Hobson, J. A., Pace-Schott, E. F., and Stickgold, R. (2000). "Dreaming and the Brain: Toward a Cognitive Neuroscience of Conscious States." *Behavioral and Brain Sciences*, *23*(06), pp. 793–842, 803.

13 Hobson, J. A., Pace-Schott, E. F., and Stickgold, R. (2000). "Dreaming and the Brain: Toward a Cognitive Neuroscience of Conscious States." *Behavioral and Brain Sciences*, 23(06), 793–842, 822.

14 Talbot, M. (2009). "Nightmare Scenario." *New Yorker*, 85(37), 42–51.

15 Buzzi, G. (2011). "False Awakenings in Light of the Dream Protoconsciousness Theory: A Study in Lucid Dreamers." *International Journal of Dream Research*, 4(2), pp. 110–116.

16 Davies, J. (2017). "Why You Shouldn't Tell people about Your Dreams." *Scientific American*, posted online May 9. Retrieved December 6, 2017 from https://blogs.scientificamerican.com/observations/why-you-shouldnt-tell-people-about-your-dreams/.

17 Symons, D. (1993). "The Stuff that Dreams Aren't Made Of: Why Wake-State and Dream-State Sensory Experiences Differ." *Cognition*, 47(3), pp. 181–217.

18 Zadra, A. L., Nielsen, T. A., and Donderi, D. C. (1998). "Prevalence of Auditory, Olfactory, and Gustatory Experiences in Home Dreams." *Percept. Mot. Skills* 87, pp. 819–826.

19 Schwartz, T. L., and Vahgei, L. (1998). "Charles Bonnet Syndrome in Children." *Journal of American Association for Pediatric Ophthalmology and Strabismus*, 2(5), pp. 310–313.

Thanks to Renaud Jardri for a conversation on this topic.

20 Cognitive activities account for 42.4 percent of waking activity, but only 18.6 percent of dreaming activity. We don't yet know why, but theories include the cholinergic state of the brain, lessened activity of the dorsolateral prefrontal cortex, and differences in hypothalamic regulation.

Schredl, M., and Hofmann (2002). "Continuity between Waking Activities and Dream Activities." *Consciousness and Cognition*, 12(2), pp. 298–308.

Hobson, J. A., Pace-Schott, E. F., and Stickgold, R. (2000). "Dreaming and the Brain: Toward a Cognitive Neuroscience of Conscious States. *Behavioral and Brain Sciences*, 23(06), pp. 793–842, 883.

21 Hobson, J. A., Pace-Schott, E. F., and Stickgold, R. (2000). "Dreaming and the Brain: Toward a Cognitive Neuroscience of Conscious States." *Behavioral and Brain Sciences*, 23(06), pp. 793–842, 805.
22 Hobson, J. A., Pace-Schott, E. F., and Stickgold, R. (2000). "Dreaming and the Brain: Toward a Cognitive Neuroscience of Conscious States." *Behavioral and Brain Sciences*, 23(06), pp. 793–842, 806.
23 These experiments are described in: LeDoux, J. (1996). *The Emotional Brain: The Mysterious Underpinnings of Emotional Life.* New York: Simon & Schuster. pp. 59.
24 These experiments are described in: LeDoux, J. (1996). *The Emotional Brain: The Mysterious Underpinnings of Emotional Life.* New York: Simon & Schuster. pp. 60.
25 Hobson, J.A., Pace-Schott, E. F. and Stickgold, R. (2000). "Dream science 2000: A Response to Commentaries on *Dreaming and the Brain.*" *Behavioral and Brain Sciences*, 23(06), pp. 1019–1035.
26 Revonsuo, A. (2000). "Did Ancestral Humans Dream for Their Lives?" *Behavioral and Brain Sciences*, 23(06), pp. 1063–1082, 1066.
27 Hobson, J. A., Pace-Schott, E. F., and Stickgold, R. (2000). "Dreaming and the Brain: Toward a Cognitive Neuroscience of Conscious States." *Behavioral and Brain Sciences*, 23(06), pp. 793–842, 826.
28 Hobson, J. A., Pace-Schott, E. F., and Stickgold, R. (2000). "Dreaming and the Brain: Toward a Cognitive Neuroscience of Conscious States." *Behavioral and Brain Sciences*, 23(06), pp. 793–842, 826.
29 Goodwyn, E. D. (2012). *The Neurobiology of the Gods: How Brain Physiology Shapes the Recurrent Imagery of Myth and Dreams.* New York: Routledge.
30 Stein, D. J. (1997). *Cognitive Science and the Unconscious.* American Psychiatric Press. pp. 104–107.
31 Hartmann, E., and Brezler, T. (2008). "A Systematic Change in Dreams after 9/11/01." *Sleep* 31, pp. 213–218.
32 Goodwyn, E. D. (2012). *The Neurobiology of the Gods: How Brain Physiology Shapes the Recurrent Imagery of Myth and Dreams.* New York: Routledge. Kindle location 1919.
33 Riveted, pp. 58–59.
34 McCarley and Hoffman, 1981.
35 Hobson, J. A., Pace-Schott, E. F., and Stickgold, R. (2000). "Dreaming and the Brain: Toward a Cognitive Neuroscience of Conscious States." *Behavioral and Brain Sciences*, 23(06), pp. 793–842, 805.
36 Revonsuo, A. (2000). "Did Ancestral Humans Dream for Their Lives? *Behavioral and Brain Sciences*, 23(06), p. 1063–1082, 1076.
37 Revonsuo, A. (2000). "The Reinterpretation of Dreams: An Evolutionary Hypothesis of the Function of Dreaming." *Behavioral and Brain Sciences*, 23(06), pp. 877–901, 897.
38 Hobson, J. A., Pace-Schott, E. F., and Stickgold, R. (2000). "Dreaming and the Brain: Toward a Cognitive Neuroscience of Conscious States." *Behavioral and Brain Sciences*, 23(06), pp. 793–842, 823.
39 Goodwyn, E. D. (2012). *The Neurobiology of the Gods: How Brain Physiology Shapes the Recurrent Imagery of Myth and Dreams.* New York: Routledge. p. 167.
40 Revonsuo, A. (2000). "Did Ancestral Humans Dream for Their Lives?" *Behavioral and Brain Sciences*, 23(06), pp. 1063–1082, 1065.
41 Hobson, J. A., Pace-Schott, E. F., and Stickgold, R. (2000). "Dreaming and the Brain: Toward a Cognitive Neuroscience of Conscious States." *Behavioral and Brain Sciences*, 23(06), pp. 793-842, 883.
Sacks, O. (2012). *Hallucinations.* New York: Alfred A. Knopf. p. 208.

42 Revonsuo, A. (2000). "Did Ancestral Humans Dream for Their Lives? *Behavioral and Brain Sciences*, 23(06), pp. 1063–1082.

43 Jones, B. E. (2000). "Dreaming as Play." *Behavioral and Brain Sciences*, 23(06), pp. 955–956.

44 Hobson, J. A., Pace-Schott, E. F., and Stickgold, R. (2000). "Dreaming and the Brain: Toward a Cognitive Neuroscience of Conscious States." *Behavioral and Brain Sciences*, 23(06), pp. 793–842, 825.

45 Hobson, J. A., Pace-Schott, E. F., and Stickgold, R. (2000). "Dreaming and the Brain: Toward a Cognitive Neuroscience of Conscious States." *Behavioral and Brain Sciences*, 23(06), pp. 793–842, 823.

46 Especially the brain area called Brodmann's area forty. See: Hobson, J. A., Pace-Schott, E. F., and Stickgold, R. (2000). "Dreaming and the Brain: Toward a Cognitive Neuroscience of Conscious States." *Behavioral and Brain Sciences*, 23(06), pp. 793–842, 826.

47 Hobson, J. A., Pace-Schott, E. F., and Stickgold, R. (2000). "Dreaming and the Brain: Toward a Cognitive Neuroscience of Conscious States." *Behavioral and Brain Sciences*, 23(06), pp. 793–842, 833.

48 Hobson, J. A., Pace-Schott, E. F., and Stickgold, R. (2000). "Dreaming and the Brain: Toward a Cognitive Neuroscience of Conscious States." *Behavioral and Brain Sciences*, 23(06), pp. 793–842, 832.

49 Voss, U. (2014). "Unlocking the Lucid Dream." *Scientific American*, winter, special issue on creativity. pp. 66–69.

50 LaBerge, S. (2000). "Lucid Dreaming; Evidence and Methodology." *Behavioral and Brain Sciences*, 23(06), pp. 962–964.

51 Walker, M. (2017). *Why we sleep: Unlocking the power of sleep and dreams*. New York: Simon and Schuster.

52 Hobson, J. A., Pace-Schott, E. F., and Stickgold, R. (2000). "Dreaming and the Brain: Toward a Cognitive Neuroscience of Conscious States." *Behavioral and Brain Sciences*, 23(06), pp. 793–842, 837.

53 Voss, U. (2014). "Unlocking the Lucid Dream." *Scientific American*, winter, special issue on creativity. pp. 66–69.

54 Voss, U. (2014). "Unlocking the Lucid Dream." *Scientific American*, winter, special issue on creativity. pp. 66–69.

55 Stumbrys, T., Erlacher, D., and Schmidt, S. (2011). "Lucid Dream Mathematics: An Explorative Online Study of Arithmetic Abilities of Dream Characters." *International Journal of Dream Research*, 4(1), pp. 35–40.

56 Tholey, P. (1989). "Consciousness and Abilities of Dream Char-Acters Observed during Lucid Dreaming." Perceptual and Motor Skills, 68(2), pp. 567–78.

57 Stumbrys, T., and Daniels, M. (2010). "An Exploratory Study of Creative Problem Solving in Lucid Dreams: Preliminary Findings and Methodological Considerations." *International Journal of Dream Research*, 3(2), pp. 121–129.

58 Rated 9.1 out of a twenty-eight-point checklist, reliably rated by two raters. This paper has a good overview of all the recommended methods, for those hell-bent on lucid dreaming: Stumbrys, T., Erlacher, D., Schädlich, M., and Schredl, M. (2012). "Induction of Lucid Dreams: A Systematic Review of Evidence." *Consciousness and Cognition*, 21(3), pp. 1456–1475.

59 I did write a science fiction play called "Read it Twice . . . It's Real!" The title is based on this reality testing. Email me if you want to produce it!

60 Naiman, R. (2017). "Dreamless: The Silent Epidemic of REM Sleep Loss." *Annals of the New York Academy of Sciences*. 1406, pp. 77–85.

61 Sacks, O. (2012). *Hallucinations*. New York: Alfred A. Knopf. p. 42.

Naiman, R. (2017). "Dreamless: The Silent Epidemic of REM Sleep Loss. *Annals of the New York Academy of Sciences*. 1406, pp. 77–85.

62 Flanagan, O. (1995). *Sleep, Dreams, and the Evolution of the Conscious Mind*. Oxford: Oxford University Press.

Revonsuo, A. (2000). "The Reinterpretation of Dreams: An Evolutionary Hypothesis of the Function of Dreaming. *Behavioral and Brain Sciences*, 23(06), pp. 877–901, 880.

63 Support for memory consolidation in dreaming: Hobson, J. A., Pace-Schott, E. F., and Stickgold, R. (2000). "Dream Science 2000: A Response to Commentaries on *Dreaming and the Brain*." *Behavioral and Brain Sciences*, 23(06), pp. 1019–1035.

Challenges: Vertes, R. P., and Eastman, K. E. (2000). "The Case against Memory Consolidation in REM Sleep." *Behavioral and Brain Sciences*, 23, pp. 867–876.

64 Kramer, M. (1993) "The Selective Mood Regulatory Function of Dreaming: An Update and Revision." In A. Moffitt, M. Kramer, and R. Hoffman (Eds.). *The Functions of Dreaming*. Albany: State University of New York Press.

65 Cartwright, R. D. (1996) "Dreams and Adaptation to Divorce." In: *Trauma and Dreams*, ed. D. Barrett. Harvard University Press.

66 Hobson, J. A., Pace-Schott, E. F., and Stickgold, R. (2000). "Dreaming and the Brain: Toward a Cognitive Neuroscience of Conscious States." *Behavioral and Brain Sciences*, 23(06), pp. 793–842, 883–4.

67 Hobson, J. A., Pace-Schott, E. F., and Stickgold, R. (2000). "Dreaming and the Brain: Toward a Cognitive Neuroscience of Conscious States." *Behavioral and Brain Sciences*, 23(06), pp. 793–842, 886.

68 Hobson, J. A., Pace-Schott, E. F., and Stickgold, R. (2000). "Dreaming and the Brain: Toward a Cognitive Neuroscience of Conscious States. *Behavioral and Brain Sciences*, 23(06), pp. 793–842, 877–901.

69 Goodwyn suggests that there is an "innate imagery" in the human mind that is dominant during unconscious imagination, such as dreaming. He cites the commonality of certain dream tropes as evidence. Goodwyn, E. D. (2012). *The Neurobiology of the Gods: How Brain Physiology Shapes the Recurrent Imagery of Myth and Dreams*. New York: Routledge.

70 Revonsuo, A. (2000). "The Reinterpretation of Dreams: An Evolutionary Hypothesis of the Function of Dreaming." *Behavioral and Brain Sciences*, 23(06), pp. 877–901, 891.

Some think that Revonsuo's view of the evolutionary environment is too bleak, and cite evidence that many contemporary hunter gatherers (such as the Kalahari Bushmen) lead a unstressful life with little danger. See, for example, Humphrey, N. (2000). "Dreaming as Play." *Behavioral and Brain Sciences*, 23(06), p. 953.

71 Revonsuo, A. (2000). "Did Ancestral Humans Dream for Their Lives?" *Behavioral and Brain Sciences*, 23(06), pp. 1063–1082, 1071.

72 Revonsuo, A. (2000). "Did Ancestral Humans Dream for Their Lives?" *Behavioral and Brain Sciences*, 23(06), pp. 1063–1082, 1070.

73 Hobson, J. A., Pace-Schott, E. F., and Stickgold, R. (2000). "Dreaming and the Brain: Toward a Cognitive Neuroscience of Conscious States." *Behavioral and Brain Sciences*, 23(06), pp. 793–842, 885.

74 Cook, M., and Mineka, S. (1990). "Selective Associations in the Observational Conditioning of Fear in Rhesus Monkeys." *Journal of Experimental Psychology: Animal Behavior Processes*, 16(4), pp. 372–389.

75 Revonsuo, A. (2000). "Did Ancestral Humans Dream for their Lives?" *Behavioral and Brain Sciences*, 23(06), pp. 1063–1082, 1070.

76 Hobson, J. A., Pace-Schott, E. F., and Stickgold, R. (2000). "Dreaming and the Brain: Toward a Cognitive Neuroscience of Conscious States." *Behavioral and Brain Sciences*, 23(06), pp. 793–842, 885.

77 Revonsuo, A. (2000). "The Reinterpretation of Dreams: An Evolutionary Hypothesis of the Function of Dreaming." *Behavioral and Brain Sciences*, 23(06), pp. 877–901, 886.

78 Revonsuo, A. (2000). "Did Ancestral Humans Dream for Their Lives?" *Behavioral and Brain Sciences*, 23(06), pp. 1063–1082.

79 Revonsuo, A. (2000). "Did Ancestral Humans Dream for Their Lives? *Behavioral and Brain Sciences*, 23(06), pp. 1063–1082, 1070

80 Revonsuo, A. (2000). "Did Ancestral Humans Dream for Their Lives? *Behavioral and Brain Sciences*, *23*(06), pp. 1063–1082, 1071.

81 Revonsuo, A. (2000). "The Reinterpretation of Dreams: An Evolutionary Hypothesis of the Function of Dreaming." *Behavioral and Brain Sciences*, 23(06), pp. 877–901, 887.

82 Revonsuo, A. (2000). "The Reinterpretation of Dreams: An Evolutionary Hypothesis of the Function of Dreaming." *Behavioral and Brain Sciences*, 23(06), pp. 877–901, 889

83 Levin, R. (2000). "Nightmares: Friend or Foe?" *Behavioral and Brain Sciences*, 23(06), p. 965.

84 Revonsuo, A. (2000). "Did Ancestral Humans Dream for Their Lives?" *Behavioral and Brain Sciences*, 23(06), pp. 1063–1082, 1072.

85 Revonsuo, A. (2000). "The Reinterpretation of Dreams: An Evolutionary Hypothesis of the Function of Dreaming." *Behavioral and Brain Sciences*, 23(06), pp. 877–901, 889.

86 Inhibitory cells in the area of the brain known as "the pons" normally cause muscle atonia. This is also the source of evidence that other mammals dream. When cats' muscle atonia is removed, they engage in species-specific behavior, such as stalking, when in REM. REM Sleep Behavior Disorder is not the same thing as sleepwalking (somnambulism), which occurs during NREM sleep. Revonsuo, A. (2000). "The Reinterpretation of Dreams: An Evolutionary Hypothesis of the Function of Dreaming. *Behavioral and Brain Sciences*, 23(06), pp. 877–901, 889–90, 892–893.

87 Revonsuo, A. (2000). "Did Ancestral Humans Dream for Their Lives?" *Behavioral and Brain Sciences*, 23(06), pp. 1063–1082, 1075.

88 McNamara, P., McLaren, D., Smith, D., Brown, A., and Stickgold, R. (2005). "A 'Jekyll and Hyde' Within: Aggressive Versus Friendly Interactions in REM and Non-REM Dreams." *Psychological Science*, 16(2), pp. 130–136.

A caveat might be in order here, however. This McNamara study found that dreaming has more social interaction than *contemporary* waking life. We don't know for sure, but it seems plausible that the majority of human evolutionary time was spent in small hunter-gatherer groups that features much more social interaction than we see in contemporary, industrialized society.

Nielsen, T. A, and Germain, A. (2000). "Post-Traumatic Nightmares as a Dysfunctional State" *Behavioral and Brain Sciences*, 23(06), pp. 978–9.

89 Revonsuo, A., Tuominen, J., and Valli, K. (2015). "The Avatars in the Machine: Dreaming as a Simulation of Social Reality." In *Open MIND*. Frankfurt am Main: MIND Group.

90 Bednar, J. A. (2000). "Internally-generated Activity, Non-Episodic Memory, and Emotional Salience in Sleep." *Behavioral and Brain Sciences*, 23(06), pp. 908–909.

Cheyne, J.A. (2000). "Play, Dreams, and Simulation." *Behavioral and Brain Sciences*, 23(06), pp. 918–919.

Humphrey, N. (2000). "Dreaming as Play. *Behavioral and Brain Sciences*, 23(06), p. 953.

Weisberg, D. S., and Gopnik, A. (2013). "Pretense, Counterfactuals, and Bayesian Causal Models: Why What is Not Real Really Matters." *Cognitive Science*, 37(7), pp. 1368–1381.

91 Hobson, J. A., Hong, C. C. H., and Friston, K. J. (2014). "Virtual Reality and Consciousness Inference in Dreaming." *Frontiers in Psychology*, 5, pp. 1–18.

92 Solms, M., personal communication, 2017.

93 Revonsuo, A. (2000). "The Reinterpretation of Dreams: An Evolutionary Hypothesis of the Function of Dreaming." *Behavioral and Brain Sciences*, 23(06), pp. 877–901, 889.

94 In the rare case of lucid dreaming, we are aware that we are dreaming, and indeed, lucid dreaming is sometimes used to treat recurring nightmares. Revonsuo, A. (2000). "The Reinterpretation of Dreams: An Evolutionary Hypothesis of the Function of Dreaming. *Behavioral and Brain Sciences*, 23(06), pp. 877–901, 889.

95 Rozen, N., and Soffer-Dudek, N. (2018). "Dreams of Teeth Falling Out: An Empirical Investigation of Physiological and Psychological Correlates." *Frontiers in Psychology*, 9.

96 Talbot, M. (2009). "Nightmare Scenario." *New Yorker,* 85(37), pp. 42–51.

97 Ardito, R. B. (2000). "Dreaming as an Active Construction of Meaning." *Behavioral and Brain Sciences*, 23(06), pp. 907–908.

7. MIND-WANDERING AND DAYDREAMING

1 Smallwood, J., McSpadden, M., and Schooler, J. W. (2008). "When Attention Matters: The Curious Incident of the Wandering Mind." *Memory & Cognition*, 36(6), pp. 1144–1150.

2 Klinger, E. (1978). "Modes of Normal Conscious Flow." In K. S. Pope, and J. L. Singer (Eds.). *The Stream of Consciousness: Scientific Investigations into the Flow of Human Experience.* New York: Plenum. pp. 225–258

3 Klinger, E. (2009). "Daydreaming and Fantastizing: Thought Flow and Motivation." In K. D. Markman, W. M. P. Klein, and J. A. Suhr (Eds.). *Handbook of Imagination and Mental Simulation.* New York: Taylor & Francis Group. pp. 225–239.

4 Glausiusz, J. (2014). "Living in an Imaginary World." *Scientific American Mind*, winter, special issue on creativity. pp. 70–77.

5 Tambini, A., Ketz, N., and Davachi, L. (2010). "Enhanced Brain Correlations during Rest Are Related to Memory for Recent Experiences." *Neuron,* 65(2), pp. 280–290.

Van de Ven, V. (2013). "Brain Functioning When the Voices Are Silent: Aberrant Default Modes in Auditory Verbal Hallucinations." In R. Jardri, A. Cachia, P. Thomas, and D. Pins (Eds.). *The Neuroscience of Hallucinations*. New York: Springer. pp. 393–415, 393, 396, 400.

6 Glausiusz, J. (2014). "Living in an Imaginary World." *Scientific American Mind*, winter, special issue on creativity. pp. 70–77. Schacter, D. L., Addis, D. R., Hassabis, D., Martin, V. C., Spreng, R. N., and Szpunar, K. K. (2012). "The Future of Memory: Remembering, Imagining, and the Brain." *Neuron*, 76(21), pp. 677–694.

7 Mason, M. F., Norton, M. I., Van Horn, J. D., Wegner, D. M., Grafton, S. T., and Macrae, C. N. (2007). "Wandering Minds: The Default Network and Stimulus-Independent Thought." *Science,* 315, pp. 393–395.

8 Future: 48 percent. Present: 29 percent. Past: 12 percent. Nontemporal thinking: 11 percent: Baird, B., Smallwood, J., and Schooler, J.W. (2011). "Back to the Future: Autobiographical Planning and the Functionality of Mind-Wandering." *Consciousness and Cognition*, 20(4), pp. 1604–1611.

9 Kroll-Mensing, D. 1992. "Differentiating Anxiety and Depression: An Experience Sampling Analysis." (Unpublished doctoral dissertation). University of Minnesota, Minneapolis, MN.

10 Glausiusz, J. (2014). "Living in an Imaginary World." *Scientific American Mind*, winter, special issue on creativity. pp. 70–77.

11 This literature is reviewed in: Mooneyham, B. W., and Schooler, J. W. (2013). "The Costs and Benefits of Mind-Wandering: A Review." *Canadian Journal of Experimental Psychology*, 67(1), pp. 11–18.

12 Andrade, J. (2010). "What Does Doodling Do?" *Applied Cognitive Psychology*, 24(1), pp. 100–106.

13 I explore this theory of boredom in this paper: Davies, J., and Fortney, M. (2012). "The Menton Theory of Boredom and Engagement." *Proceedings of the First Annual Conference on Advances in Cognitive Systems*. pp. 131–143.

14 Casner, S. M., and Schooler, J. W. (2014). "Thoughts in Flight: Automation Use and Pilots' Task-Related and Task-Unrelated Thought." *Human Factors*, 56(3), pp. 433–442.

15 Williams, J., Merritt, J., Rittenhouse, C., and Hobson, J. A. (1992). "Bizarreness in Dreams and Fantasies: Implications for the Activation-Synthesis Hypothesis." *Consciousness & Cognition*, 1, pp. 172–185.

16 Klinger, E. (2009). "Daydreaming and Fantastizing: Thought Flow and Motivation." In K. D. Markman, W. M. P. Klein, and J. A. Suhr (Eds.). *Handbook of Imagination and Mental Simulation*. New York: Taylor & Francis Group. pp. 225–239.

17 Bokhour, B. G., Clark, J. A., Inui, T. S., Sillilman, R. A., and Talcott, J. A. (2001). "Sexuality after Treatment for Early Prostate Cancer: Exploring the Meanings of 'Erectile Dysfunction.'" *Journal of General Internal Medicine*, 16(10), pp. 649–655.

18 Taylor, S. E., Pham, L. B., Rivkin, I. D., and Armor, D. A. (1998). "Harnessing the Imagination: Mental Simulation, Self-Regulation, and Coping." *American Psychologist*, 53, pp. 429–439.

19 Baird, B., Smallwood, J., Mrazek, M. D., Kam, J. W. Y., Franklin, M. S., and Schooler, J. W. (2012). "Inspired by Distraction: Mind Wandering Facilitates Creative Incubation." *Psychological Science*, 23(10), pp. 1117–1122.

20 Killingsworth, M. A., and Gilbert, D. (2010). "A Wandering Mind is an Unhappy Mind." *Science*, 330, p. 932.

Westgate, E., Wilson, T. D., Gilbert, D. T. (unpublished manuscript). "The Difficulty of Enjoying Deliberative Thought."

21 Wilson, T. D., Reinhard, D. A., Westgate, E. C., Gilbert, D. T., Ellerbeck, N., Hahn, C., and Shaked, A. (2014). "Just Think: The Challenges of the Disengaged Mind." *Science*, 345(6192), pp. 75–77.

22 Shalev, S. (2011). "Solitary Confinement and Supermax Prisons: A Human Rights and Ethical Analysis." *Journal of Forensic Psychology Practice*, 11(2–3), pp. 151–183.

23 Neilson, S. (2016). "How to Survive Solitary Confinement." *Nautilus*, Jan/Feb, pp. 59–63.

24 Glausiusz, J. (2014). "Living in an Imaginary World." *Scientific American Mind*, winter, special issue on creativity. pp. 70–77.

25 Bigelsen, J., and Schupak, C. (2011). "Compulsive Fantasy: Proposed Evidence of an Under-Reported Syndrome through a Systematic Study of 90 Self-Identified Non-Normative Fantasizers." *Consciousness and Cognition*, 20, pp. 1634–1648.

26 There is no official name for it yet. Bigelsen, J., and Schupak, C. (2011). "Compulsive Fantasy: Proposed Evidence of an Under-Reported Syndrome through a Systematic Study

of 90 Self-Identified Non-Normative Fantasizers." *Consciousness and Cognition*, 20, pp. 1634–1648.

27 Glausiusz, J. (2014). "Living in an Imaginary World." *Scientific American Mind*, winter, special issue on creativity. pp. 70—77.

28 Glausiusz, J. (2014). "Living in an Imaginary World." *Scientific American Mind*, winter, special issue on creativity. pp. 70–77.

29 Schupak, C., and Rosenthal, J. (2009). "Excessive Daydreaming: A Case History and Discussion of Mind Wandering and High Fantasy Proneness." *Consciousness and Cognition*, 18(1), pp. 290–292.

30 Killingsworth, M. A., and Gilbert, D. (2010). "A Wandering Mind is an Unhappy Mind." *Science*, 330, p. 932.

31 People rated their vacations on a scale of how relaxing they were. After the vacation, those who said their trip was "relaxed" had no difference in happiness upon coming home than people who did not go on vacation. It was only people who reported their trip as being "very relaxed" who enjoyed increased happiness upon return. This change in happiness lasted a maximum of two weeks, after which there is a return to pretrip happiness levels.

The length of the vacation had no effect: Nawijn, J., Marchand, M. A., Veenhoven, R., and Vingerhoets, A. J. (2010). "Vacationers Happier, but Most not Happier after a Holiday." *Applied Research in Quality of Life*, 5(1), pp. 35–47.

8. IMAGINATION AS MENTAL TRAINING, HEALING, AND SELF-IMPROVEMENT

1 Bain, A. (1872). *The Senses and the Intellect.* New York: D. Appleton and Co.;

Jacobson, E. (1931). "Electrical Measurement of Neuromuscular States during Mental Activities." *American Journal of Physiology*, 96, pp. 115–121.

Lotze, M., Montoya, P., Erb, M., Hulsmann, E., Flor, H, Klose, U., et al. (1999). "Activation of Cortical and Cerebellar Motor Areas during Executed and Imagined Hand Movements: An fMRI Study." *Journal of Cognitive Neuroscience*, 11(5), pp. 491–501.

2 Dijkerman, C., Ietswaart, M., and Johnston, M. (2010). "Motor Imagery and the Rehabilitation of Movement Disorders: An Overview." In *The Neurophysiological Foundations of Mental and Motor Imagery*. Oxford: Oxford University Press. pp. 127–143.

3 Fourkas, A. D., Avenanti, A., Urgesi, C., and Aglioti, S. M. (2006). "Corticospinal Facilitation during First and Third Person Imagery." *Experimental Brain Research*, 168(1-2), pp. 143–151.

4 Stinear, C. (2010). "Corticospinal Facilitation during Motor Imagery." In *The Neurophysiological Foundations of Mental and Motor Imagery*. Oxford: Oxford University Press. pp. 47–61, 58.

5 That is, the prefrontal regions, the supplementary motor area (SMA), and the cerebellum. Interestingly, there is not an overlap in activation for the primary motor cortex. Beilock, S. L., and Lyons, I. M. (2009). "Expertise and the Mental Simulation of Action." In K. D. Markman, W. M. P. Klein, and J. A. Suhr (Eds.). *Handbook of Imagination and Mental Simulation*. New York: Taylor & Francis Group. pp. 21–34.

6 Collet, C., and Guillot, A. (2010). "Autonomic Nervous System Activities during Imagined Movements." In *The Neurophysiological Foundations of Mental and Motor Imagery*. Oxford: Oxford University Press. pp. 95–107, 100.

7 Arora, S., Aggarwal, R., Sirimanna, P., Moran, A., Grantcharov, T., Kneebone, R., and Darzi, A., et al. (2011). "Mental Practice Enhances Surgical Technical Skills: A Randomized Controlled Study." *Annals of Surgery*, 253(2), pp. 265–270.

8 Kosslyn, S. M., and Moulton, S. T. (2009). "Mental Imagery and Implicit Memory." In K. D. Markman, W. M. P. Klein, and J. A. Suhr (Eds.). *Handbook of Imagination and Mental Simulation*. New York: Taylor & Francis Group. pp. 35–52.

9 This is a 30.4 percent increase from a baseline score of 5.2 successful putts for every 10 attempts.
Woolfolk, R. L., Parrish, M. W., and Murphy, S. M. (1985). "The Effects of Positive and Negative Imagery on Motor Skill Performance." *Cognitive Therapy and Research*, 9, pp. 335–341.

10 Kosslyn, S. M., and Moulton, S. T. (2009). "Mental Imagery and Implicit Memory." In K. D. Markman, W. M. P. Klein, and J. A. Suhr (Eds.). *Handbook of Imagination and Mental Simulation*. New York: Taylor & Francis Group. pp. 35–52.

11 In this case, the force went from 9.14 to 11.15, which is an increase of 22.03 percent.
Yue, G. H., and Cole, K. J. (1992). "Strength Increases from the Motor Program: Comparison of Training with Maximal Voluntary and Imagined Muscle Contractions." *Journal of Neurophysiology*, 67, pp. 1114–1123.

12 Stumbrys, T. (2015). *Motor Learning in Lucid Dreams: Prevalence, Induction, and Effectiveness* (Doctoral dissertation).

13 Driskell, J. E., Copper, C., and Moran, A. (1994). "Does Mental Practice Enhance Performance?" *Journal of Applied Psychology*, 79, pp. 481–492.

14 Shooter games are more effective than puzzle or role-playing video games.
Granic, I., Lobel, A., and Engels, R. C. M. E. (2014). "The Benefits of Playing Video Games." *American Psychologist*. 69(1), pp. 66–78.

15 Morewedge, C. K., Huh, Y. E., and Vosgerau, J. (2010). "Thought for Food: Imagined Consumption Reduces Actual Consumption." Science, 303, pp. 1530–1533.

16 Spence, C., Okajima, K., Cheok, A. D., Petit, O., and Michel, C. (2016). "Eating With our Eyes: From Visual Hunger to Digital Satiation." *Brain and Cognition*, 110, pp. 53–63.

17 Bower, G. H. (1970). "Analysis of a Mnemonic Device: Modern Psychology Uncovers the Powerful Components of an Ancient System for Improving Memory." *American Scientist*, 58(5), pp. 496–510.

18 Nairne, J. S. (2010). "Adaptive Memory: Evolutionary Constraints on Remembering." *Psychology of Learning and Motivation*, 53, pp. 1–32.

19 Zwaan, R. A., Van den Broek, P., Truitt, T. P., and Sundermeier, B. (1996). "Causal Coherence and the Accessibility of Object Locations in Narrative Comprehension." *Abstracts of the Psychonomic Society*, 1, p. 50.

20 Note that even these associations have mnemonics. The 2 and 3 look like sideways n and m, respectively. See the Wikipedia page on the mnemonic major system for more.

21 The part of the brain called the "parahippocampus" is implicated in memory and imagination of travel and place locations. Tversky, B. (2005). "Functional Significance of Visuospatial Representations." In P. Shah and A. Miyake (Eds.). *The Cambridge Handbook of Visuospatial Thinking*. Cambridge: Cambridge University Press. p. 10.
You can hear me interviewed about memory palaces on the excellent podcast "The Reality Check." http://www.trcpodcast.com/trc-505-memory-palaces-donating-used-clothing/

22 Bower, G. H. (1970). "Analysis of a Mnemonic Device: Modern Psychology Uncovers the Powerful Components of an Ancient System for Improving Memory." *American Scientist*, 58(5), pp. 496–510.

23 Einstein, G. O., McDaniel, M. A., and Lackey, S. (1989). "Bizarre Imagery, Interference, and Distinctiveness." *Journal of Experimental Psychology: Learning, Memory, and Cognition*, 15(1), p. 137.

24 http://blog.ted.com/diana-nyad-epic-playlist/.

25 Maddison, R., Prapavessis, H., Clatworthy, M., Hall, C., Foley, L., Harper, T., Cupal, D., and Brewer, B. (2012). "Guided Imagery to Improve Functional Outcomes Post-Anterior Cruciate Ligament Repair: Randomized-Controlled Pilot Trial." *Scandinavian Journal of Medicine & Science in Sports*, 22(6), pp. 816–821.

26 Dickstein, R., and Deutsch, J. E. (2001). "Motor Imagery in Physical Therapy Practice." *Physical Therapy*, 87, pp. 942–953.

27 Pincus, D., and Sheikh, A. A. (2009). *Imagery for Pain Relief: A Scientifically Grounded Guidebook for Clinicians.* New York: Routledge. pp. 67—68.

28 Pincus, D., and Sheikh, A. A. (2009). *Imagery for Pain Relief: A Scientifically Grounded Guidebook for Clinicians.* New York: Routledge. pp. 62–63, 67–68.

29 Pincus, D., and Sheikh, A.A. (2009). *Imagery for Pain Relief: A Scientifically Grounded Guidebook for Clinicians.* New York: Routledge. p. 81.

30 Pincus, D., and Sheikh, A. A. (2009). *Imagery for Pain Relief: A Scientifically Grounded Guidebook for Clinicians.* New York: Routledge. p. 72.

It is worth mentioning that belief in the effectiveness of a treatment has, indeed, been shown to create greater effectiveness, but even people who don't believe it can get benefits from placebo or sham treatments, as shown in this study of acupuncture: Linde, K., Witt, C. M., Streng, A., Weidenhammer, W., Wagenpfeil, S., Brinkhaus, B., Melchart, D., et al. (2007). "The Impact of Patient Expectations on Outcomes in Four Randomized Controlled Trials of Acupuncture in Patients with Chronic Pain." *Pain*, 128(3), pp. 264–271.

31 Pincus, D., and Sheikh, A. A. (2009). *Imagery for Pain Relief: A Scientifically Grounded Guidebook for Clinicians.* New York: Routledge. pp. 234–138.

32 Pincus, D., and Sheikh, A. A. (2009). *Imagery for Pain Relief: A Scientifically Grounded Guidebook for Clinicians.* New York: Routledge. pp. 158.

33 Davies, J. (2014). *Riveted: The Science of Why Jokes Make Us Laugh, Movies Make Us Cry, and Religion Makes Us Feel One with the Universe.* New York: Palgrave Macmillan.

34 Byrne, B., Rollag, M. D., Hanifin, J. P., Reed, C., and Brainard, G. C. (2000). "Bright Light Imagery Does not Suppress Melatonin." *Journal of Pineal Research*, 29(1), pp. 62–64.

35 This is called incompatible emotive imagery. For a review of different therapeutic imagery types, see: Fernandez, E. (1986). "A Classification System for Cognitive Coping Strategies for Pain." *Pain*, 26, pp. 141–151.

36 This kind of imagery is known as stimulus transformative imagery: Pincus, D., and Sheikh, A. A. (2009). *Imagery for Pain Relief: A Scientifically Grounded Guidebook for Clinicians.* New York: Routledge. pp. 74.

Fernandez, E. (1986). "A Classification System for Cognitive Coping Strategies for Pain." *Pain*, 26, pp. 141–151.

37 Spanos, N. P., Horton, C., and Chaves, J. R. (1975). "The Effects of Two Cognitive Strategies on Pain Threshold." *Journal of Abnormal Psychology*, 85, pp. 677–681.

38 Pincus, D., and Sheikh, A. A. (2009). *Imagery for Pain Relief: A Scientifically Grounded Guidebook for Clinicians.* New York: Routledge. pp. 163.

39 Pincus, D., and Sheikh, A. A. (2009). *Imagery for Pain Relief: A Scientifically Grounded Guidebook for Clinicians.* New York: Routledge. pp. 177.

40 Pincus, D., and Sheikh, A. A. (2009). *Imagery for Pain Relief: A Scientifically Grounded Guidebook for Clinicians.* New York: Routledge. pp. 145, 161, 184.

41 Pincus, D., and Sheikh, A. A. (2009). *Imagery for Pain Relief: A Scientifically Grounded Guidebook for Clinicians.* New York: Routledge. p. 146.

42 Pincus, D., and Sheikh, A. A. (2009). *Imagery for Pain Relief: A Scientifically Grounded Guidebook for Clinicians.* New York: Routledge. pp. 147 and 149.

43 Pincus, D., and Sheikh, A. A. (2009). *Imagery for Pain Relief: A Scientifically Grounded Guidebook for Clinicians.* New York: Routledge. pp. 146 and 148.

44 Knox, V. J. (1972). "Cognitive Strategies for Coping with Pain: Ignoring vs. Acknowledging." Unpublished doctoral dissertation, University of Waterloo, Waterloo.

45 Lazarus, A. (1977). *In the Mind's Eye.* New York: Rawson Associates. pp. 21–27.

46 I started with this book, which teaches mantra meditation:

Easwaran, E. (2008). *Passage Meditation: Bringing the Deep Wisdom of the Heart into Daily Life.* Nilgiri Press.

I also highly recommend Hanh, T. N. (2002). *Teachings on Love.* Berkeley: Parallax Press.

47 Davies, J. (2016). "You Can Have Emotions You Don't Feel." *Nautilus*, August 17, blog entry.

48 Dutton, D. G., and Aron, A. P. (1974). "Some Evidence for Heightened Sexual Attraction under Conditions of High Anxiety. *Journal of Personality and Social Psychology*, 30(4), p. 510.

49 Dawson, M. (2016). "Unraveling the Secret Science of Falling in Love." *New York Post,* May 21, retrieved from http://nypost.com/2016/05/21/unraveling-the-secret-science-of-falling-in-love/. July 14, 2017.

50 Sze, J. A., Gyurak, A., Yuan, J. W., and Levenson, R. W. (2010). "Coherence between Emotional Experience and Physiology: Does Body Awareness Training have an Impact?" *Emotion*, 10(6), pp. 803–814.

51 Winkielman, P., and Berridge, K. C. (2004). "Unconscious Emotion." *Current Directions in Psychological Science*," 13(3), pp. 120–123.

52 Williams, L. M., Liddell, B. J., Kemp, A. H., Bryant, R. A., Meares, R. A., Peduto, A. S., and Gordon, E. (2006). "Amygdala–Prefrontal Dissociation of Subliminal and Supraliminal Fear." *Human Brain Mapping*, 27(8), pp. 652–661.

53 Petre, M., and Blackwell, A. F. (1999). "Mental Imagery in Program Design and Visual Programming." *International Journal of Human-Computer Studies*, 51, pp. 7–30.

54 Retrieved from https://hiphopdx.com/news/id.20809/title.kanye-west-says-he-didnt-physically-write-his-lyrics-for-first-four-albums#.

55 LeBoutillier, N., and Marks, D. F. (2003). "Mental Imagery and Creativity: A Meta-Analytic Review Study." *British Journal of Psychology*, 94(1), pp. 29–44.

56 Reality monitoring is one of the most important metacognitive tasks people perform. Johnson, M. K., and Raye, C. L. (1981). "Reality Monitoring." *Psychological Review,* 88, pp. 67–85.

57 Tetlock, P., and Belkin, A. (1996). *Counterfactual Thought Experiments in World Politics: Logical, Methodological, and Psychological Perspectives.* Princeton, NJ: Princeton University Press.

58 This study has been done many times; 23 percent is the result from: Frank, M. C., and Ramscar, M. (2003). "How do Presentation and Context Influence Representation for Functional Fixedness Tasks?" In *Proceedings of the 25th Annual Meeting of the Cognitive Science Society.* p. 1345.

59 Galinsky, A. D., and Moskowitz, G. B. (2000). "Counterfactuals as Behavioral Primes: Priming the Simulation Heuristic and Consideration of Alternatives." *Journal of Experimental Social Psychology,* 36, pp. 257–383.

60 Markman, K. D., Lindberg, M. J., Kray, L. J., and Galinsky, A. D. (2007). Implications of Counterfactual Structure for Creative Generation and Analytical Problem Solving." *Personality and Social Psychology Bulletin*, 33, pp. 312–324.

61 Kray, L. J., Galinsky, A. D., and Wong, E. M. (2006). "Thinking Inside the Box: The Rational Processing Style Elicited by Counterfactual Mind-Sets." *Journal of Personality and Social Psychology*, 91, pp. 33–48.

62 Wong, E. M., Galinsky, A. D., and Kray, L. J. (2009). The Counterfactual Mind-Set: A Decade of Research." In K. D. Markman, W. M. P. Klein, and J. A. Suhr (Eds.). *Handbook of Imagination and Mental Simulation.* New York, NY: Taylor & Francis Group. pp. 161–174

63 Shepard, R. N. (1978). "The Mental Image." *American Psychologist*, *33*(2), p. 125. Nersessian, N. J. (1992). "How Do Scientists Think? Capturing the Dynamics of Conceptual Change in Science." *Cognitive Models of Science*, 15, pp. 3–44.

64 Nersessian, N. J. (2005). "Abstraction via Generic Modeling in Concept Formation in Science. *Idealization XII: Correcting the Model. Idealization and Abstraction in the Sciences (Poznan Studies in the Philosophy of the Sciences and the Humanities, Vol. 86).* Rodopi, Amsterdam, pp. 117–143.

65 Uttal, D. H., Meadow, N. G., Tipton, E., Hand, L. L., Alden, A. R., Warren, C., and Newcombe, N. S. (2013). "The Malleability of Spatial Skills: A Meta-Analysis of Training Studies. *Psychological Bulletin*, 139(2), p. 352.

66 The fortress/tumor problem is from: Duncker, K. (1926). "A Qualitative (Experimental and Theoretical) Study of Productive Thinking (Solving of Comprehensible Problems). *The Pedagogical Seminary and Journal of Genetic Psychology*, 33(4), pp. 642–708.

67 I modeled this analogy using artificial intelligence: Davies, J., and Goel, A. K. (2001). "Visual Analogy in Problem Solving." *Proceedings of the International Joint Conference on Artificial Intelligence*. Morgan Kaufmann Publishers. pp. 377–382.

68 Dennett, D.C. (2013). *Intuition Pumps and Other Tools for Thinking.* New York: W. W. Norton. pp. 99.

69 Dawkins, R. (2005). "Why the Universe Seems so Strange." Talk at *TED*, July. https://www.ted.com/talks/richard_dawkins_on_our_queer_universe?language=en#t-44135.

9. IMAGINARY COMPANIONS

1 Taylor, M., Hodges, S. D., and Kohanyi, A. (2003). "The Illusion of Independent Agency: Do Adult Fiction Writers Experience their Characters as having Minds of their Own?" *Imagination, Cognition, and Personality*, 22, pp. 361–38.

2 Taylor, M. (1999). *Imaginary Companions and the Children who Create Them.* New York: Oxford University Press. p. 113.

3 Taylor, M. (1999). *Imaginary Companions and the Children Who Create Them.* New York: Oxford University Press. p. 11.

4 Taylor, M. (1999). *Imaginary Companions and the Children Who Create Them.* New York: Oxford University Press. p. 11.

5 Taylor, M. (1999). *Imaginary Companions and the Children Who Create Them.* New York: Oxford University Press. p. 9.

6 Taylor, M. (1999). *Imaginary Companions and the Children Who Create Them.* New York: Oxford University Press. p. 12.

7 Taylor, M. (1999). *Imaginary Companions and the Children Who Create Them.* New York: Oxford University Press. p. 21.

8 Taylor, M. (1999). *Imaginary Companions and the Children Who Create Them.* New York: Oxford University Press. p. 23.

9 Taylor, M. (1999). *Imaginary Companions and the Children Who Create Them.* New York: Oxford University Press. p. 14.

10 Taylor, M. (1999). *Imaginary Companions and the Children Who Create Them.* New York: Oxford University Press. p. 15.

11 Taylor, M. (1999). *Imaginary Companions and the Children Who Create Them.* New York: Oxford University Press. p. 17.

12 Taylor, M. (1999). *Imaginary Companions and the Children Who Create Them.* New York: Oxford University Press. p. 9.

13 Taylor, M. (1999). *Imaginary Companions and the Children Who Create Them.* New York: Oxford University Press. p. 128.

14 Taylor, M. (1999). *Imaginary Companions and the Children Who Create Them.* New York: Oxford University Press. p. 118.

15 Taylor, M. (1999). *Imaginary Companions and the Children Who Create Them.* New York: Oxford University Press. p. 123.

16 Taylor, M. (1999). *Imaginary Companions and the Children Who Create Them.* New York: Oxford University Press. p. 131.

17 Taylor, M. (1999). *Imaginary Companions and the Children Who Create Them.* New York: Oxford University Press. p. 136.

18 Taylor, M. (1999). *Imaginary Companions and the Children Who Create Them.* New York: Oxford University Press. p. 140.

19 This might be related to the ancient Greek idea of the role of muses in the creative process. Taylor, M. (1999). *Imaginary Companions and the Children Who Create Them.* New York: Oxford University Press. p. 148.

For a book on this subject see: Watkins, M. (1990). *Invisible Guests: The Development of Imaginal Dialogues.* Boston: Sigo Press.

20 Watkins, M. (2000). *Invisible Guests: The Development of Imaginal Dialogues.* Woodstock, CT: Spring Press. pp. 95–101.

The panel discussion was at World Fantasy Con, November 5, 2015, on "The Rogue." Panelists were Ellen Kushner (moderator), Linda Williams Chima, James Alan Gardner, Alistair Kimble, and Ryk E. Spoor.

21 Taylor, M., Hodges, S. D., and Kohanyi, A. (2003). "The Illusion of Independent Agency: Do Adult Fiction Writers Experience their Characters as Having Minds of Their Own?" *Imagination, Cognition and Personality,* 22, pp. 361–368.

22 Taylor, M., Hodges, S. D., and Kohanyi, A. (2003). "The Illusion of Independent Agency: Do Adult Fiction Writers Experience Their Characters as Having Minds of Their Own? *Imagination, Cognition, and Personality*, 22, pp. 361–38.

23 Foxwell, J. (2018). "'I Won't be Involved with this Fictional Plot': Characters' Agency and Authors' Intentions." Talk at *Personification Across Disciplines*, Durham, UK.

24 Foxwell, J. (2018). "'I Won't be Involved with this Fictional Plot': Characters' Agency and Authors' Intentions." Talk at *Personification Across Disciplines*, Durham, UK.

25 This is based on the author's own self-reporting, so part of this could be that writers remember their imaginary friends better, perhaps because their imagination is more vivid, or their imaginary companions are more memorable. Green, M. C., and Donahue, J. K. (2009). "Simulated Worlds: Transportation into Narratives. In K. D. Markman, W. M. P. Klein, and J. A. Suhr (Eds.). *Handbook of Imagination and Mental Simulation.* New York: Taylor & Francis Group. pp. 241–254.

26 Carson, S. (2014). "The Unleashed Mind." *Scientific American Mind*, Winter, special issue on creativity. pp. 28–35.

27 Taylor, M. (1999). *Imaginary Companions and the Children Who Create Them.* New York: Oxford University Press. p. 18.

28 Harris, P. L. (2000). *The Work of the Imagination*. Oxford: Blackwell Publishers Ltd. p. 58.

29 Taylor, M. (1999). *Imaginary Companions and the Children Who Create Them*. New York: Oxford University Press. p. 19.

30 Taylor, M. (1999). *Imaginary Companions and the Children Who Create Them*. New York: Oxford University Press. p. 127.

31 Another possibility is that the intuition of something being rational (or not) is the result of a general pattern matching, rather than an automatized reasoning process. That is, something might feel irrational because it appears to be much like other irrational things experienced in the past. When we try to articulate *why* it's irrational, we hope that deliberate reasoning supports the conclusion reached by our pattern matching.

32 Though I might be the first to suggest that imaginary companions and autonomous fictional characters are the same kind of thing, Mary Watkins hints at the idea on page 99. Watkins, M. (2000). *Invisible Guests: The Development of Imaginal Dialogues*. Woodstock, CT: Spring Press.

33 Taylor, M. (1999). *Imaginary Companions and the Children Who Create Them*. New York: Oxford University Press. p. 152.

34 Stanovich, K. E. (2004). *The Robot's Rebellion*. Chicago: The University of Chicago Press.

35 The Tibetans use it differently. A tulpa of a spider might be created to get over a fear of spiders, but then it would be abandoned. Tulpamancers have a forum on Tulpa.info.

Veissière, S. (2016) "Varieties of Tulpa Experiences: The Hypnotic Nature of Human Sociality, Personhood, and Interphenomenality." In A. Raz and M. (Eds). *Hypnosis and Meditation: Towards an Integrative Science of Conscious Planes*. Oxford: Oxford University Press. pp. 55–76.

36 22.5 percent of tulpamancers think there is more than a psychological explanation for tulpas.

Veissière, S. (2016) "Varieties of Tulpa Experiences: The Hypnotic Nature of Human Sociality, Personhood, and Interphenomenality." In A. Raz and M. (Eds). *Hypnosis and Meditation: Towards an Integrative Science of Conscious Planes*. Oxford: Oxford University Press. pp. 55–76.

37 Veissière, S. (2016) "Varieties of Tulpa Experiences: The Hypnotic Nature of Human Sociality, Personhood, and Interphenomenality." In A. Raz and M. (Eds). *Hypnosis and Meditation: Towards an Integrative Science of Conscious Planes*. Oxford: Oxford University Press. pp. 55–76, 62, 68.

38 Luhrmann, T. M. (2004). "Metakinesis: How God becomes Intimate in Contemporary US Christianity." *American Anthropologist*, 106(3), pp. 518–528.

39 Davis, P., personal communication. 2018.

I discussed this theory with imaginary companion researcher Paige Davis at a 2018 conference. With respect to childhood imaginary companions, there isn't yet data on whether or not they start out autonomous.

40 Taylor, M. (1999). *Imaginary Companions and the Children Who Create Them*. New York: Oxford University Press. p. 4.

41 Taylor, M. (1999). *Imaginary Companions and the Children Who Create Them*. New York: Oxford University Press. Chapter 3.

42 Taylor, M. (1999). *Imaginary Companions and the Children Who Create Them*. New York: Oxford University Press. p. 37.

43 Harris, P. L. (2000). *The Work of the Imagination*. Oxford: Blackwell Publishers Ltd.

44 Taylor, M. (1999). *Imaginary Companions and the Children Who Create Them*. New York: Oxford University Press. p. 112.

45 Taylor, M. (1999). *Imaginary Companions and the Children Who Create Them*. New York: Oxford University Press. p. 89.
46 Taylor, M. (1999). *Imaginary Companions and the Children Who Create Them*. New York: Oxford University Press. p. 90.
47 Taylor, M. (1999). *Imaginary Companions and the Children Who Create Them*. New York: Oxford University Press. p. 98.
48 Taylor, M. (1999). *Imaginary Companions and the Children Who Create Them*. New York: Oxford University Press. p. 88.
49 Taylor, M., and Mottweiler, C. M. (2008). "Imaginary Companions: Pretending They Are Real but Knowing They Are Not. *American Journal of Play*, summer, pp. 47–54.
50 Taylor, M. (1999). *Imaginary Companions and the Children Who Create Them*. New York: Oxford University Press. p. 90.
51 Taylor, M. (1999). *Imaginary Companions and the Children Who Create Them*. New York: Oxford University Press. pp. 91–92.
52 Taylor, M. (1999). *Imaginary Companions and the Children Who Create Them*. New York: Oxford University Press. p. 91.
Davies, J. (2018). "The Science of Saying Goodbye to Santa." *The Conversation*. Published online December 17.
53 Taylor, M. (1999). *Imaginary Companions and the Children Who Create Them*. New York: Oxford University Press. pp. 95–96.
54 Taylor, M. (1999). *Imaginary Companions and the Children Who Create Them*. New York: Oxford University Press. pp. 55–56.
Davies, J. (2014). "Is Santa Claus a God?" *Nautilus*, December 24 blog entry.
55 Taylor, M. (1999). *Imaginary Companions and the Children Who Create Them*. New York: Oxford University Press. p. 56.
56 Taylor, M. (1999). *Imaginary Companions and the Children Who Create Them*. New York: Oxford University Press. pp. 115–116.
57 Taylor, M. (1999). *Imaginary Companions and the Children Who Create Them*. New York: Oxford University Press. pp. 62, 76, 77.
58 Taylor, M. (1999). *Imaginary Companions and the Children Who Create Them*. New York: Oxford University Press. p. 6.
59 Taylor, M. (1999). *Imaginary Companions and the Children Who Create Them*. New York: Oxford University Press. p. 82.
60 Taylor, M. (1999). *Imaginary Companions and the Children Who Create Them*. New York: Oxford University Press. p. 82.
61 Taylor, M. (1999). *Imaginary Companions and the Children Who Create Them*. New York: Oxford University Press. p. 84.
62 Taylor, M. (1999). *Imaginary Companions and the Children Who Create Them*. New York: Oxford University Press. pp. 67–70.
63 Taylor, M. (1999). *Imaginary Companions and the Children Who Create Them*. New York: Oxford University Press. p. 64.
64 Taylor, M. (1999). *Imaginary Companions and the Children Who Create Them*. New York: Oxford University Press. p. 57.
65 Taylor, M. (1999). *Imaginary Companions and the Children Who Create Them*. New York: Oxford University Press. p. 51.
66 Taylor, M. (1999). *Imaginary Companions and the Children Who Create Them*. New York: Oxford University Press. p. 53.

67 Wenner Moyer, M. (2014). "The Serious Need for Play." *Scientific American Mind* special issue on creativity, Winter, pp. 78–85.

68 Wenner Moyer, M. (2014). "The Serious Need for Play." *Scientific American Mind* special issue on creativity, Winter, pp. 78–85.

69 Wallace, C. E., and Russ, S. W. (2015). "Pretend Play, Divergent Thinking, and Math Achievement in Girls: A Longitudinal Study." *Psychology of Aesthetics, Creativity, and the Arts*, 9(3), pp. 296–305.

70 Taylor, M. (1999). *Imaginary Companions and the Children Who Create Them*. New York: Oxford University Press. p. 158.

10. IMAGINATION AND TECHNOLOGY

1 Stokes, M., Thompson, R., Cusack, R., and Duncan, J. (2009). "Top-Down Activation of Shape-Specific Population Codes in Visual Cortex during Mental Imagery." *Journal of Neuroscience*, 29(5), pp. 1565–1572.

2 Cowen, A. S., Chun, M. M., and Kuhl, B. A. (2014). "Neural Portraits of Perception: Reconstructing Face Images from Evoked Brain Activity." *Neuroimage*, 94, pp. 12--22.

3 Kay, K. N., Naselaris, T., Prenger, R. J., and Gallant, J. L. (2008). "Identifying Natural Images from Human Brain Activity." *Nature*, 452, pp. 352–355.

Chadwick, M. J., Hassabis, D., Weiskopf, N., and Maguire, E. A. (2010). "Decoding Individual Episodic Memory Traces in the Human Hippocampus." *Current Biology*, 20, pp. 544–547.

4 Seeliger, K., Güçlü, U., Ambrogioni, L., Güçlütürk, Y., and van Gerven, M. A. J. (2018). "Generative Adversarial Networks for Reconstructing Natural Images from Brain Activity." *NeuroImage*, 181, pp. 775–785.

5 Morton, G. A. (1962, 2000). "Machines with Imagination." *Proceedings of the IEEE*, 88(2), p. 283.

6 Tversky, B., and Hemenway, K. (1983). "Categories of Environmental Scenes." *Cognitive Psychology*, 15(1), pp. 121–149.

7 Tversky, B., and Hemenway, K. (1983). "Categories of Environmental Scenes." *Cognitive Psychology*, 15(1), pp. 121–149.

8 Von Ahn, L., and Dabbish, L. (2004). "Labeling Images with a Computer Game." In *Proceedings of the SIGCHI Conference on Human Factors in Computing Systems*. ACM. pp. 319–326.

9 http://www.scholarpedia.org/article/Visual_short_term_memory.

10 Vertolli, M. O., Kelly, M. A., and Davies, J. (2018). "Coherence in the Visual Imagination." *Cognitive Science*. 43(3): pp. 885–917.

11 Technically, we use a mode, not a mean angle. This is because sometimes the mean angle is meaningless. Take, for example, the relationship between nose and an ear. The ear is almost always to the left of or to the right of the nose (say, 90 degrees and 270 degrees). If it's right of 50 percent of the time, then the mean angle would be above or below! So we would take, in this example, either the 90 or the 270.

12 Chowdhury, W., Akkaoui, A., Burt, C., and Davies, J. (2015). "Quanty: An Online Game for Eliciting the Wisdom of the Crowd." *Computers in Human Behavior*. 49, pp. 213–219.

13 Technically, for this part of the process, we use information from a different, but similar database called LabelMe. LabelMe is made by volunteers who outlined objects in pictures. But because it does not require agreement between volunteers, it is subject to people putting in bad data, like designating an arbitrary outline and calling it an airplane. But we use it for the demonstration because, when it is accurate, it has better and more complete outlines than

Peekaboom, which only tends to highlight the most distinguishing features of objects: for an airplane, for example, players learned that the tail was the most distinguishing feature, so the pixels that were supposed to represent an airplane were actually just the pictures of airplanes' tails. A similar problem happens with the label "person" and giving pixels just for a person's face.

Von Ahn, L., Liu, R., and Blum, M. (2006). "Peekaboom: A Game for Locating Objects in Images." In *Proceedings of the SIGCHI Conference on Human Factors in Computing Systems*. ACM. pp. 55–64.

Russell, B. C., Torralba, A., Murphy, K., Freeman, W., (2008), LabelMe: a database and web-based tool for image annotation, *International Journal of Computer Vision*, vol. 77(1-3).

14 Kosslyn, S. M., Thompson, W. L., and Ganis, G. (2006). *The Case for Mental Imagery*. New York: Oxford University Press. p. 43.

15 Morris, C. (2011, March 5). "World of Warcraft Maker Turns 20, Looks Ahead." Plugged In. Retrieved October 3, 2011 from Plugged In:http://games.yahoo.com/blogs/plugged-in/world-warcraft-maker-turns-20-looks-ahead-450.html.

16 http://aigamedev.com/broadcasts/sessions-ssx3/.

17 http://mewo2.com/notes/terrain/.

You can also follow it on twitter, and get random fantasy maps in your feed: @unchartedatlas.

18 Breault, V. (2015). "Multi-Agent Planning for Interactive Emergent Narrative Systems." Unpublished master's thesis. Carleton University, Ottawa.

19 Veale, T. (2017). "Déjà Vu All Over Again: On the Creative Value of Familiar Elements in the Telling of Original Tales." In Proceedings of *ICCC 2017, the 8th International Conference on Computational Creativity, Atlanta, Georgia*.

20 http://www.escapistmagazine.com/news/view/114126-Skyrim-Is-a-Neverending-Story.

21 Weiner, J. (2011). "Where Do Dwarf-Eating Carp Come From? *New York Times Magazine*. July 21.

http://www.nytimes.com/2011/07/24/magazine/the-brilliance-of-dwarf-fortress.html.

22 Hall, C. (2014). "Dwarf Fortress Will Crush Your CPU Because Creating History Is Hard." Polygon, July 23, retrieved August 24, 2017 from: https://www.polygon.com/2014/7/23/5926447/dwarf-fortress-will-crush-your-cpu-because-creating-history-is-hard.

23 Imagination is discussed as a form of virtual reality in: Hobson, J. A., Hong, C. C. H., and Friston, K. J. (2014). "Virtual Reality and Consciousness Inference in Dreaming." *Frontiers in Psychology*, 5.

Dreams were compared to virtual reality in: Revonsuo, A., and Salmivalli, C. (1995). "A Content Analysis of Bizarre Elements in Dreams." *Dreaming*, 5(3), p. 169.

24 Stapleton, C., and Davies, J. (2011). "Imagination: The Third Reality to the Virtuality Continuum." *2011 IEEE International Symposium on Mixed and Augmented Reality*. (ISMAR-2011). Basel, Switzerland. pp. 53–60.

11. HOW IMAGINATION WORKS

1 Woodworth, R. S. (1906). "Imageless Thought." *Journal of Philosophy, Psychology and Scientific Methods*, 3, pp. 701–708, 703.

2 Richardson, A. (1983). "Imagery: Definition and Types." In A. A. Sheikh (Ed.). *Imagery: Current Theory, Research, and Application*. New York: Wiley and Sons. pp. 3–42, 6.

3 Some have tried to classify types of mental imagery. One classification has afterimages, eidetic images, thought images, and imagination images.

Richardson, A. (1983). "Imagery: Definition and Types." In A. A. Sheikh (Ed.), *Imagery: Current Theory, Research, and Applications.* New York: Wiley. pp. 3–42.

The simplest form is the afterimage, which is what you sometimes see when you close your eyes after looking at the sun or a camera flash. It is an aftereffect of a strong or prolonged stimulus. Like other forms of imagery, they need not be visual. You can have an afterimage of something stuck in your throat after it's gone, or the feeling that your shirt is still on after it's taken off.

Eidetic images are a bit like afterimages but are more stable and come from memory. If an eidetic child looks at a picture, after that picture is taken away, they sometimes claim to still be able to "see" the picture there, even when they move their eyes. These kinds of images are hard to relate to because very few children and almost no adults ever experience them, though they might appear in therapeutic techniques that guide imagery.

Haber, R. N., and Haber, L. (2000). "Eidetic Imagery." In A. E. Kazdin (Ed.). *Encyclopedia of Psychology,* vol. 3, pp. 147–149.

Thought imagery is when you imagine something in response to an external stimulus, such as when you picture something you're reading or when you imagine what someone sounds like when their voice is raw.

Imagination imagery is like thought imagery, but with a deep focus on the images and a deliberate attempt to *not* focus on external stimulation. These categories are not cut-and-dried, and a given instance of imagery might well be in between two of these classes.

4 One might consider each pixel to be a symbol, and the pixel and its location to be a proposition, but this is stretching pretty far what people consider to be symbols and propositions. In general, symbols are meaningful, and pixels are not. Only collections of pixels, arranged in certain ways, get *perceived* as meaningful—and then symbols about them are recognized or created. In cognitive science, pixels are considered to be sub-symbolic, which kind of means that they don't represent anything meaningful in the real world, but rather parts of meaningful things.

5 This neural tissue need not be spatially contiguous in the brain. That is, the neuron clusters representing different "pixels" need not be next to each other in the brain—they just need to be functionally connected in a spatial configuration, even though it might be distributed over a larger part of the brain that's also doing other things.

6 Kosslyn, S. M., Thompson, W. L., and Ganis, G. (2006). *The Case for Mental Imagery.* New York: Oxford University Press. p. 179.

7 Reisberg, D., and Heuer, F. (2005). "Visuospatial Images." In P. Shah and A. Miyake (Eds.). *The Cambridge Handbook of Visuospatial Thinking.* Cambridge: Cambridge University Press. p. 48.

8 Reisberg, D., and Heuer, F. (2005). "Visuospatial Images." In P. Shah and A. Miyake (Eds.). *The Cambridge Handbook of Visuospatial Thinking.* Cambridge: Cambridge University Press. pp. 35–80.

9 Kosslyn, S. M., Thompson, W. L., and Ganis, G. (2006). *The Case for Mental Imagery.* New York: Oxford University Press. p. 87.

10 Kosslyn, S. M., Thompson, W. L., and Ganis, G. (2006). *The Case for Mental Imagery.* New York: Oxford University Press. p. 83.

11 Kosslyn, S. M., Thompson, W. L., and Ganis, G. (2006). *The Case for Mental Imagery.* New York: Oxford University Press. p. 164.

12 Kosslyn, S. M., Thompson, W. L., and Ganis, G. (2006). *The Case for Mental Imagery.* New York: Oxford University Press. p. 172.

13 Damage to inferior temporal lobes, which process visual information, disrupts visualization abilities but not spatial imagery, and damage to the posterior parietal lobes, which process spatial information, has the opposite effect. Blind people are slower at mental scanning between locations in an imagined scene.

Kosslyn, S. M., Thompson, W. L., and Ganis, G. (2006). *The Case for Mental Imagery.* New York: Oxford University Press. pp. 143, 164.

14 Visual tasks seem to use the occipital lobe (particularly areas V1 and V2) and the ventral stream in the brain, where spatial tasks seem to use the temporal lobe and the ventral stream of processing. Reisberg, D., and Heuer, F. (2005). "Visuospatial Images." In P. Shah and A. Miyake (Eds.). *The Cambridge Handbook of Visuospatial Thinking.* Cambridge, UK: Cambridge University Press. pp. 35–80.

15 This is related to object permanence, which is the ability to know that something has not ceased to exist after it has left your immediate sensory awareness.

Baird, A. A., Kagan, J., Gaudette, T., Walz, K. A., Hershlag, N., and Boas, D. A. (2002). "Frontal Lobe Activation during Object Permanence: Data from Near-Infrared Spectroscopy." *NeuroImage*, 16(4), pp. 1120–1126.

16 Attneave, F., and Farrar, P. (1977). "The Visual World behind the Head." *The American Journal of Psychology*, pp. 549–563, 561.

17 One study found that people could reinterpret rotated figures that they only perceived by touch! It could be that they are rotating a haptic (touch) image, or it could be that the haptic information is contributing to the spatial image.

Pouw, W., Aslanidou, A., Kamermans, K., and Paas, F. (2017). "Is Ambiguity Detection in Haptic Imagery Possible? Evidence for Enactive Imaginings." In G. Gunzelmann, A. Howes, T. Tenbrink, and E. J. Davelaar (Eds.). *Proceedings of the 39th Annual Conference of the Cognitive Science Society.* Austin, TX: Cognitive Science Society. pp. 2925–2930.

18 Kosslyn, S. M., Thompson, W. L., and Ganis, G. (2006). *The Case for Mental Imagery.* Oxford: Oxford University Press. p. 167.

19 Faw, B. (1997). "Outlining a Brain Model of Mental Imaging Abilities." *Neuroscience & Biobehavioral Reviews*, 21(3), pp. 283–288.

Faw, B. (2009). "Conflicting Intuitions May be Based on Differing Abilities: Evidence from Mental Imaging Research." *Journal of Consciousness Studies*,16(4), pp. 45–68.

This is also suggested by philosopher Jesse Prinz in an interview on the *Philosophy Bites* podcast "Jesse Prinz on Thinking with Pictures." http://philosophybites.libsyn.com/jesse-prinz-on-thinking-with-pictures.

20 These three classes were called MI-0, MI-1, and MI-2 and introduced by: Faw, B. (1997). "Outlining a Brain Model of Mental Imaging Abilities." *Neuroscience & Biobehavioral Reviews*, 21(3), pp. 283–288.

21 Stoerig, P., and Cowey, A. (1997). "Blindsight in Man and Monkey." *Brain*, 120(3), pp. 535–559.

The similarity of blindsight to nonconscious imagery was also introduced in: Faw, B. (1997). "Outlining a Brain Model of Mental Imaging Abilities." *Neuroscience & Biobehavioral Reviews*, 21(3), pp. 283–288.

22 To use more scientific jargon, blindsight patients can see sharp onset/offset times, motion, and low spatial frequency. Sharp onset/offset times: Alexander, I., and Cowey, A. (2010). "Edges, Colour and Awareness in Blindsight." *Consciousness and Cognition*, 19(2), pp. 520–533.

Motion: Ffytche, D. H., and Zeki, S. (2010). "The Primary Visual Cortex, and Feedback to It, Are Not Necessary for Conscious Vision." *Brain*, 134(1), pp. 247–257.

Low spatial frequency: Sahraie, A., Hibbard, P. B., Trevethan, C. T., Ritchie, K. L., and Weiskrantz, L. (2010). Consciousness of the first order in blindsight. *Proceedings of the National Academy of Sciences, 107*(49), pp. 21217–21222.

23 D'Angiulli, A. (2002). "Mental Image Generation and the Contrast Sensitivity Function." *Cognition*, 85(1), pp. B11–B19.

24 Schwitzgebel, E. (2002). "How Well Do We Know Our Own Conscious Experience? The Case of Visual Imagery." *Journal of Consciousness Studies*, 9(5–6), pp. 3553.

25 Salthouse, T. A., Babcock, R. L., Skovronek, E., Mitchell, D. R., and Palmon, R. (1990). "Age and Experience Effects in Spatial Visualization." *Developmental Psychology*, 26(1), p. 128.

26 Also, imagery interferes with perception more when lying down. Sheikh, A. A. (2002). *Handbook of Therapeutic Imagery Techniques.* Amityville, NY: Baywood Publishing Company. p. 386.

27 Galton, F. (1883). *Inquiries into Human Faculty and its Development.* London: Macmillan.

28 Titus, S., and Horsman, E. (2009) "Characterizing and Improving Spatial Visualization Skills." *Journal of Geoscience Education*, 57(4), pp. 242–254.

29 Ben-Chaim, D., Lappan, G., and Houang, R. T. (1988). "The Effect of Instruction on Spatial Visualization Skills of Middle School Boys and Girls." *American Educational Research Journal*, 25(1), pp. 51–71.

30 Pérez-Fabello, M. J., and Campos, A. (2007). "Influence of Training in Artistic Skills on Mental Imaging Capacity." *Creativity Research Journal*, 19(2–3), pp. 227–232.

Parrot C. A. (1986) "Visual Imagery Training: Stimulating Utilization of Imaginal Processes." *Journal of Mental Imagery*, 10, pp. 47–64.

There is some suggestive evidence that motor imagery can improve with practice, too: Nobbe, J. A., Nilsen, D., and Gillen, G. (2012). "Mental Practice Is Modifiable: Changes in Perspective Maintenance and Vividness Post-Stroke." *Journal of Imagery Research in Sport and Physical Activity*, 7(1). pp. 2–16.

31 Abumrad, J., and Krulwich, R. (2012). "Seeing in the Dark." *RadioLab* podcast episode, Oct 22. http://www.radiolab.org/story/245482-seeing-dark/.

32 Nobbe, J. A., Nilsen, D., and Giller, G. (2012). "Mental Practice Is Modifiable: Changes in Perspective Maintenance and Vividness Post-Stroke." *Journal of Imagery Research in Sport and Physical Activity*, 7:1, pp. 1932–0191.

33 The following paper was cited in another paper as finding that people with low vividness had a marked improvement after training, but there were no changes for the high-vividness group. I cannot find the paper to verify this, however: Walsh, F. J., White, K. D., and Ashton, R. (1978). "Imagery Training: Development of a Procedure and Its Evaluation." Unpublished Research Report, University of Queensland.

The University of Queensland was nice enough to go looking for this paper for me, but unfortunately their records don't go back that far. They even contacted Ashton, who said that the paper copy was lost with a bunch of other things in the flood of Brisbane. If you find a copy of this paper, email it to me!

34 Sommer's book has an entire chapter titled "Visualization Training" that asserts confidently that practice can improve your vividness and control of imagery, but cites no studies that support the claim. This is particularly striking because he does cite empirical studies that support other claims in the chapter, such as that activities such as regular breathing and lying down help support imagery. So you'd think he'd have cited evidence if there were any. It's an old book, but the others I've found are similar.

Sommer, R. (1978). *The Mind's Eye: Imagery in Everyday Life*. Palo Alto, CA: Dale Seymour Publications.

35 Lazarus, A. (1977). *In the Mind's Eye*. New York: Rawson Associates. p. 44.

36 Rademaker, R. L., and Pearson, J. (2012). "Training Visual Imagery: Improvements of Metacognition, but Not Imagery Strength." *Mental Imagery*, 23.

37 Richardson, A., and Patterson, Y. (1986). "An Evaluation of Three Procedures for Increasing Imagery Vividness." In A. A. Sheikh (Ed.). *International Review of Mental Imagery*, vol. 2, Human Sciences Press, Inc. pp. 166–191.

38 Peyroux, E., and Franck, N. (2013). "An Epistemological Approach: History of Concepts and Ideas about Hallucinations in Classical Psychiatry." In R. Jardri, A. Cachia, P. Thomas, and D. Pins (Eds.). *The Neuroscience of Hallucinations*. New York: Springer. pp. 3–20, 9.

39 Stephane, M., Folstein, M., Matthew, E., and Hill, T. C. (2000). "Imaging Auditory Verbal Hallucinations during Their Occurrence." *The Journal of Neuropsychiatry and Clinical Neurosciences*, 12(2), pp. 286–287.

40 Johns, L. C., Hemsley, D., and Kuipers, E. (2002). "A Comparison of Auditory Hallucinations in a Psychiatric and Nonpsychiatric Group." *The British Journal of Clinical Psychology*, 41, pp. 81–86.

41 Ford, J. M., and Hofman, R. E. (2013). "Functional Brain Imaging of Auditory Hallucinations: From Self-Monitoring Deficits to Co-Opted Neural Resources." In R. Jardri, A. Cachia, P. Thomas, and D. Pins (Eds.). *The Neuroscience of Hallucinations*. New York: Springer. pp. 359–373, 366.

42 Garety, P. A., Kuipers, E., Fowler, D., Freeman, D., and Bebbington, P. E. (2001). "A Cognitive Model of the Positive Symptoms of Psychosis." *Psychological Medicine*, 31, pp. 189–195, 190.

43 Hill, K., and Linden, D. E. J. (2013). "Hallucinatory Experiences in Nonclinical Populations." In R. Jardri, A. Cachia, P. Thomas, and D. Pins (Eds.). *The Neuroscience of Hallucinations*. New York: Springer. pp. 21–1, 31–32.

44 Jardri, R., and Denève, S. (2013b). "Circular Inferences in Schizophrenia." *Brain, 136*, pp. 3227–3241.

45 Windt, J., and Metzinger (2007). "The Philosophy of Dreaming and Self-Consciousness: What Happens to the Experiential Subject during the Dream State?" In D. Barrett and P. McNamara (Eds.). *The New Science of Dreaming*. Westport, CT: Praeger. pp. 193–247.

46 Fénelon, G. (2013). "Hallucinations Associated with Neurological Disorders and Sensory Loss." In R. Jardri, A. Cachia, P. Thomas, and D. Pins (Eds.). *The Neuroscience of Hallucinations*. New York: Springer. pp. 59–83, 69.

47 Lewis-Williams JD (1988). The signs of all times: entoptic phenomena in Upper Palaeolithic art. *Current Anthropology*. 29 (2): 201–45.

48 Massoud, S. (2013). Standardized Assessment of Hallucinations. In R. Jardri, A. Cachia, P. Thomas, and D. Pins (Eds.). *The Neuroscience of Hallucinations*. New York: Springer. pp. 85–104, 89.

49 Hill, K., and Linden, D. E. J. (2013). "Hallucinatory Experiences in Non-Clinical Populations." In R. Jardri, A. Cachia, P. Thomas, and D. Pins (Eds.). *The Neuroscience of Hallucinations*. New York: Springer. pp. 21–41, 38.

50 Massoud, S. (2013). "Standardized Assessment of Hallucinations." In R. Jardri, A. Cachia, P. Thomas, and D. Pins (Eds.). *The Neuroscience of Hallucinations*. New York: Springer. pp. 85–104, 89.

51 Bergmann, J., Genç, E., Kohler, A., Singer, W., and Pearson, J. (2016). "Smaller Primary Visual Cortex Is Associated with Stronger, but Less Precise Mental Imagery." *Cerebral Cortex*, 26(9), pp. 3838–3850.

52 Massoud, S. (2013). "Standardized Assessment of Hallucinations." In R. Jardri, A. Cachia, P. Thomas, and D. Pins (Eds.). *The Neuroscience of Hallucinations*. New York: Springer. pp. 85–104, 89.

53 Massoud, S. (2013). "Standardized Assessment of Hallucinations." In R. Jardri, A. Cachia, P. Thomas, and D. Pins (Eds.). *The Neuroscience of Hallucinations*. New York: Springer. pp. 85–104, 91–92.

54 Massoud, S. (2013). "Standardized Assessment of Hallucinations." In R. Jardri, A. Cachia, P. Thomas, and D. Pins (Eds.). *The Neuroscience of Hallucinations*. New York: Springer. pp. 85–104, 88.

55 We know that semantic memory is important for imagery because patients with semantic memory problems have trouble imagining future episodes.

Schacter, D. L., Addis, D. R., Hassabis, D., Martin, C. C., Spreng, R. N., and Szpunar, K. K. (2012). "The Future of Memory: Remembering, Imagining, and the Brain." *Neuron*, 76(21), pp. 677–694.

56 Johnson-Laird, P. N. (1995). Mental models, deductive reasoning, and the brain. "*The Cognitive Neurosciences*, 65, pp. 999–1008.

57 Interestingly, this experiment failed to find evidence of mental imagery: Langston, W. L., Kramer, D. C., and Glenberg, A. M. (1998). "The Representation of Space in Mental Models Derived from Text." *Memory & Cognition*, 26(2), pp. 247–262.

58 These are called "descriptions" by Kosslyn, "spatial frameworks" by Tversky.

Kosslyn, S. M. (1994) *Image and Brain: The Resolution of the Imagery Debate*. Cambridge, MA: MIT Press.

Bryant, D. J., Tversky, B., and Franklin, N. (1992) "Internal and External Spatial Frameworks for Representing described Scenes." *Journal of Memory & Language*, 1, pp. 74–98.

59 Tversky, B. (2005). "Functional Significance of Visuospatial Representations." In P. Shah and A. Miyake (Eds.). *The Cambridge Handbook of Visuospatial Thinking*. Cambridge, UK: Cambridge University Press. p. 3.

60 General structure retrieved first: Aginsky, V., and Tarr, M. J. (2000). "How Are Different Properties of a Scene Encoded in Visual Memory?" *Visual Cognition*, 7(1/2/3), pp. 147–162.

Details fleshed out later: Simons, D. J., and Levin, D. T. (1997). "Change Blindness." *Trends in Cognitive Sciences*. 1(7), pp. 261–267.

61 Why am I asking you to focus on your favorite color? Because I've found that when you're talking about imagination with people, and particularly talking about how it works, they tend to get antsy about two things: 1) that imagination might not be as magical as they thought it was, and 2) they are afraid that they will appear to imagine things much like everybody else does. Because neither of these two things are very pleasant for many people, they like to thwart the point I'm trying to make by imagining unusual versions of things I ask them to imagine. For example, if I ask them to imagine a bird, they might imagine an unusual bird, like an ostrich or a dodo, because they anticipate what most people would say and react differently. By asking you to focus on the color, I'm subtly tricking you into *not* focusing on the orientation of the shape, which is the point of the exercise (color has nothing to do with the point I'm making, but people feel they can express their individuality through color choice, and let their choices about the shape of the triangle go to a default). Get you with your guard down!

62 Prototype theory: Rosch, E. (1975). "Cognitive Representation of Semantic Categories." *Journal of Experimental Psychology,* 104, pp. 192–233.

Prototypes in mathematical concepts: Armstrong, S. L., Gleitman, L. R., and Gleitman, H. (1983). What some concepts might not be. *Cognition,* 13(3), pp. 263–308.

Prototypes are processed faster: Winkielman, P., Halberstadt, J., Fazendeiro, T, and Catty, S. (2006). *Psychological Science,* 17(9), pp. 799–806.

63 Lakoff, G., and Johnson, M. (1999). *Philosophy in the Flesh.* New York: Basic Books.

64 Richardson, D. C., Spivey, M. J., Edelman, S., and Naples, A. J. (2001). "Language Is Spatial": Experimental Evidence for Image Schemas of Concrete and Abstract Verbs. In *Proceedings of the Twenty-third Annual Meeting of the Cognitive Science Society,* Erlbaum. pp. 873–878.

65 Davies, J. (2014). *Riveted: The Science of Why Jokes Make Us Laugh, Movies Make Us Cry, and Religion Makes Us Feel One with the Universe.* New York: St. Martin's Press. p. 190.

66 Tversky, B. (2005). "Functional Significance of Visuospatial Representations." In P. Shah and A. Miyake (Eds.). *The Cambridge Handbook of Visuospatial Thinking.* Cambridge, UK: Cambridge University Press. p. 19.

67 Tversky, B., Kugelmass, S., and Winter, A. (1991). "Cross-Cultural and Developmental Trends in Graphic Productions." *Cognitive Psychology,* 23(4), pp. 515–557.

68 Kosslyn, S. M., Thompson, W. L., and Ganis, G. (2006). *The Case for Mental Imagery.* Oxford: Oxford University Press. pp. 118–120.

69 Kosslyn, S. M., Thompson, W. L., and Ganis, G. (2006). *The Case for Mental Imagery.* Oxford: Oxford University Press. pp. 128, 137.

70 Kosslyn, S. M., Thompson, W. L., Kim, I. J., and Alpert, N. M. (1995). "Topographical Representations of Mental Images in Primary Visual Cortex." *Nature,* 378(30), pp. 496–498.

71 Kosslyn, S. M., and Schwartz, S. P. (1977). "A Simulation of Visual Imagery. *Cognitive Science,* 1, pp. 265–295.

Funt, B. V. (1980). "Problem-solving with Diagrammatic Representations." *Artificial Intelligence,* 13(3), pp. 201–230.

72 Oswald, I. (1957). "After-Images from Retina and Brain." *Quarterly Journal of Experimental Psychology,* 9(2), pp. 88–100.

73 Kosslyn, S. M., Thompson, W. L., and Ganis, G. (2006). *The Case for Mental Imagery.* New York: Oxford University Press. p. 71.

74 See page 58 of: Mast, F. W., and Kosslyn, S. M. (2002). "Visual Mental Images can be Ambiguous: Insights from Individual Differences in Spatial Transformation Abilities." *Cognition,* 86, pp. 57–70.

People have also found evidence of people reinterpreting haptic (touch) imagery: Pouw, W., Aslanidou, A., Kamermans, K., and Paas, F. (2017). "Is Ambiguity Detection in Haptic Imagery Possible? Evidence for Enactive Imaginings." In G. Gunzelmann, A. Howes, T. Tenbrink, and E. J. Davelaar (Eds.). *Proceedings of the 39th Annual Conference of the Cognitive Science Society* (pp. 2925–2930). Austin, TX: Cognitive Science Society.

75 Kosslyn, S. M., Thompson, W. L., and Ganis, G. (2006). *The Case for Mental Imagery.* New York: Oxford University Press. p. 84.

76 Kosslyn, S. M., Thompson, W. L., and Ganis, G. (2006). *The Case for Mental Imagery.* New York: Oxford University Press. p. 42.

77 Kosslyn, S. M., Thompson, W. L., and Ganis, G. (2006). *The Case for Mental Imagery.* New York: Oxford University Press. p. 42.

78 Kosslyn, S. M., Thompson, W. L., and Ganis, G. (2006). *The Case for Mental Imagery.* New York: Oxford University Press. p. 81.

79 Hasson, U., Chen, J., and Honey, C. J. (2015). "Hierarchical Process Memory: Memory as an Integral Component of Information Processing." *Trends in Cognitive Sciences*, 19(6), pp. 304–313.

80 Pearson, J., Clifford, C. W., and Tong, F. (2008). "The Functional Impact of Mental Imagery on Conscious Perception." *Current Biology*, 18(13), pp. 982–986.

81 Specifically, these possible paths are represented in CA3, a part of the hippocampus.
Johnson, A., and Redish, A. D. (2007). "Neural Ensembles in CA3 Transiently Encode Paths forward of the Animal at a Decision Point." *Journal of Neuroscience*, 27(45), pp. 12176–12189.

INDEX

9/11 attacks, 130–131

A

Abstract concepts, 4–14, 21–22, 36, 75–84, 168–171, 255–265
Achievements, 70–71, 85, 155, 190
Act of will, 2–3, 7, 248
Activation-Inputs-Mode (AIM) model, 135
Activation-synthesis hypothesis, 132–134, 149
Adaptation, 55–58, 125, 141, 149, 244
Addiction, 157, 168
Affective forecasting, 72
Afterimage, 92, 247, 252, 263
Against Malaria Foundation, 72
Aggression, 141–143, 168
Al-Shahristani, Hussain, 157
Alcohol consumption, 95, 121, 139, 158
Alien abduction, 108
Alien Hand Syndrome, 115, 117
Altruism, 75–76
Aminergic, 135
Amit, Elinor, 7, 80
Amnesia, 59, 126
Amputee, 119–120
Analogy, 11, 142, 190–192, 232–233
Anderson, Laurie, 3
Anderson, Michael, 13
Andrade, Jackie, 152
Angels, 90, 198
Anger, 43, 111, 140, 183–184, 201, 210
Angles, 34, 102, 219–220, 258
Animals
  brain of, 12–15
  dreaming, 123–124, 139–140
  dreams about, 130–134, 141–144
  hallucinations about, 100
  illusions of, 235–236, 264
  imagining, 218
  magnetic field and, 35
  relationships of, 4
  stuffed animals, 196–197, 208, 212
  using imagination, 267
  vision of, 193
Animation, 99, 131–132, 256–257
Antarctica, 94, 173
Antidepressants, 139
Anton-Babinski syndrome, 240–241
Anxiety, 86, 134, 139–141, 147–152, 160–164, 181
Aphantasia, 16–19, 36, 240–243, 262
Ardito, Rita, 149
Art, 88–89, 121, 125, 185–186, 225, 244, 250–251. *See also* Creativity
Artane, 98–99
Artificial intelligence, 52, 67, 214
Artificial light, 139
Atheist, 198
Athletes, 137, 148
Auditory association cortex, 96–97, 118
Auditory hallucinations, 90, 96–97, 109, 248
Auditory imagination, 15–17, 105, 194–195, 248
Augmented reality, 2, 93, 195, 227–228
Authors, 187, 199–205, 246–247
Auto-correct, 25, 28, 30
Automatization, 202–205, 246
Autonomous, 137, 159, 199–205, 246
Autonomous Set of Systems, 204
Autonomy, 199–205
Autoscopic hallucinations, 98
Autoscopy, 98
Availability heuristic, 128
Ayahuasca, 90, 99

B

Bain, Alexander, 165
Baird, Benjamin, 152, 155–156
Bandura, Albert, 68
Barber, Theodore, 92
Basketball, 158, 167

Bateson, William, 191–192
Bathroom, 9, 64, 132
Bear, 129–130
Beastie Boys, 4, 36, 174, 216
Beautiful Mind, A, 99
Beauty, 40
Behavioral programs, 141
Behaviorism, 229–231
Ben Chaim, David, 244
Bereavement, 72, 91–93, 119
Bexton, William, 109
Bibliotherapy, 39–40
Bicknell, Jeanette, 17, 49, 52
Biederman, Irving, 34
Bigelsen, Jayne, 157
Bizarre, 128–131, 154, 172
Blair, Irene, 84
Blakemore, Sarah-Jayne, 115–116
Blindness, 34–36, 73, 106, 119, 126, 195, 238–244
Blindsight, 240–241
Blouin-Hudon, Eve-Marie, 14, 68, 164
Bokhour, Barbara, 155
Boredom, 151–153, 158, 212
Boroditsky, Lera, 10
Bottom-up process, 24, 109–112
Brain stem, 12–13, 132–135, 148–149
Breault, Vincent, 225
Broca's area, 118
Brock, Timothy, 88
Buchez, Philippe, 248
Buddha, 107
Buddhism, 97, 162
Bugs Bunny, 41–42

C
California, living in, 60
Candle problem, 188
Cannabis, 139
Capgras syndrome, 95–96
CAPTCHA, 216
Carlson, Elizabeth, 39–41
Cartesian theater, 101–102
Cartwright, Rosalind, 140
Caruso, Eugene, 72, 78–79
Casner, Stephen, 153
Chadwick, Martin, 214
Chaotic information, 56
Chaotic input, 132–134, 148–151
Character, 96, 104, 108, 118
Characters, 96, 137, 143–145, 199–209, 225, 246–251
Charity, 72–75, 83
Charles Bonnet syndrome, 100–110, 126, 133, 205, 249
Childhood amnesia, 126
Children
  amnesia in, 126
  charity and, 83
  cultural differences and, 211–212
  dreams of, 123–124, 136, 142–144
  imaginary companions of, 91, 196–212
  learning processes of, 223–224, 256
  multicultural activities of, 89
  pretend play, 36–53, 92, 196–197, 201, 207–208, 212
  Santa Claus and, 208–209
  starving children, 73
  stuffed animals and, 196–197, 208, 212
  Tooth Fairy and, 208–209
Cholinergic, 135
Chowdhury, Wahida, 220
Chronicles of Narnia, 49–51
Circadian rhythms, 139
Clarity, 40, 251
Clark, Cindy Dell, 209
Clayton, Nicola, 59
Color
  dreams and, 139
  hallucinations and, 98–102, 111–113, 120–121
  imagery and, 13, 236–239, 262–265
  pixels, 232–233, 262–263
  processing, 13, 22–27, 35, 214, 257
Color Purple, The, 200
Complex hallucinations, 96–103, 109–110, 250–251
Complex objects, 1, 33–35, 236
Complex perceptions
  complex objects and, 33–35, 236
  dreams and, 127–134
  feelings and, 86–87
  imagination and, 167, 192, 261–267
  learning procedures, 197–206
  memory and, 13, 20–26
  technology and, 225
Complexity, 21, 109, 146, 206, 247, 250–251
Compulsive fantasizing, 157–163, 182
Computer vision, 220–221
Concrete concepts, 10–15, 68, 74–83, 168–172, 215, 256–259
Congenitally blind people, 36, 126, 195, 238–239
Conscious imagery, 74–75, 240–243, 247, 253

Consciousness, 15, 106, 128, 135, 138, 185, 248, 257
Continuity hypothesis, 127
Control
- of characters, 199–209
- of dreams, 132–137
- of emotions, 71–74, 210–211
- experimental control, 74, 177
- external control, 115
- feelings of, 39–40
- of hallucinations, 111–115
- of imagery, 102–103, 159–161
- of imagination, 92, 201–202, 247–251
- mood control, 88
- motor control, 11–15
- voluntary control, 247–251

Conway, Paul, 83
Copperfield, David, 53
Cornelissen, Gert, 84
Corollary discharge, 118
Costumes, 100–101, 212
Counterfactual thinking, 85–87, 187–190
Cowen, Alan, 214
Creativity, 1–2, 27, 88, 126–128, 137, 155–161, 185–190, 195, 200–212, 226, 247
Crichton, Michael, 55
Csikszentmihalyi, Mihaly, 88
Cultural differences, 211–212
Curare, 117

D

Dali, Salvador, 186
Dancers, 98, 184
D'Angiulli, Amedeo, 242
D'Argembeau, Arnaud, 65
Dawkins, Richard, 193–194
Daydreaming
- description of, 150–152
- doodling and, 152–154
- dreams and, 60, 97, 132–135, 150–164
- external stimulations, 156–161
- imagery and, 152, 159, 163
- internal stimulations, 156–161
- maladaptive daydreaming, 157–160
- meditation and, 162–164
- mind-wandering and, 71, 150–164
- mindfulness and, 161–164
- purpose of, 154–156

Deafness, 119
Death, 43, 46, 72–74, 91, 142, 173
Deep character, 202, 205
Default mode network, 151
Default network, 60, 71, 151
Dein, Simon, 107
Delusions, 44, 47, 93–101, 104, 112–119, 249
Dementia, 96, 101, 108
Demons, 90, 107–108, 198
Dennett, Daniel, 101–102, 191–192
Deontology, 80, 84
Depersonalization, 138
Depictive representations, 232–234, 239–242, 264, 267
Depression, 39–40, 59, 68–69, 140, 157, 168, 179, 182, 225
Derealization, 138
Detail, 14–19, 48, 65, 101–106, 178, 215, 229, 244–252. *See also* Imagery
Determinism, 79, 80, 221
Diabetes, 68
Diablo, 224
Diagnostic and Statistical Manual of Mental Disorders (DSM-V), 158
Diet, 83, 168
Dieting, 168
Disney, 9, 41–42, 152, 176
Dissociative identity disorder, 39, 96, 210–211
Distance, 7–9, 32, 81, 180, 219–221, 258
DMT, 93, 98, 99
DNA, 55, 190–192
Dolphins, 16, 122
Domhoff, G. William, 126
Doodling, 152–154
Dorsolateral prefrontal cortex (DLPFC), 129, 136
Dracula, 51, 54
Dreambank, 126
Dreaming. *See also* Dreams
- daydreaming and, 60, 97, 132–135, 150–164
- imagination and, 4, 15–16, 122–149, 186–187, 205, 208, 213, 227, 243, 247, 250–252
- reasons for, 139–141
- social simulation theory, 145, 149
- threat simulation theory, 141–149

Dreams. *See also* Dreaming
- about animals, 130–134, 141–144
- by animals, 123–124, 139–140
- by children, 123–124, 136, 142–144
- color and, 139
- complex perceptions and, 127–134
- content of, 126–132
- control of, 132–137
- daydreams and, 60, 97, 132–135, 150–164

fantasies and, 143, 157–164
flying dreams, 145, 186–187
hallucinations and, 47, 90, 93, 100, 108, 120
illusions and, 147, 159
imagery and, 126–145
interpretation of, 129–137, 148, 249–250
lucid dreams, 135–139, 159, 167, 200, 205, 250
meaning of, 249–250
mirroring reality, 146–149
nightmares, 39, 43, 129, 140–148, 254
reasons for, 139–141
remembering, 123–125
sexual dreams, 136, 145, 155
theories about, 141–149
Drugs, 90–99, 103, 106, 117, 121, 148, 230–231, 250
Duck/rabbit illusion, 235–236, 264. *See also* Illusions
Duncker candle problem, 188
Dungeons & Dragons, 199
Dunn, Elizabeth, 73–74
Duolingo, 216
Duplicates, 95
Dutton, Donald, 183
Dwarf Fortress, 225–226
Dysfunctional source monitoring, 91

E
Effective altruism, 75–76
Efference copy, 118
Electrical discharge, 99
Electromagnetic field theory, 190
Emmons, Robert, 87
Emotions
anger, 43, 111, 140, 183–184, 201, 210
anxiety, 86, 134, 139–141, 147–152, 160–164, 181
brain and, 13–14
complex perceptions and, 86–87
control of, 71–74, 210–211
creating, 181–182
depression, 39–40, 59, 68–69, 140, 157, 168, 179, 182, 225
empathy, 79, 89
excitation, 110–114
grief, 72, 91–93, 119
hallucinations and, 109–111
happiness, 60–63, 84–89, 155, 162–163, 179–183
imagery and, 82–87, 182–186
imagination and, 77–89, 182–186
moods and, 43, 46, 86–88, 121, 179, 185
understanding, 182–186
Empathy, 79, 89
Environment of evolutionary adaptation, 141
Epilepsy, 96, 99, 103, 108
Episodic memory, 36–39, 44, 56–59, 254
Erdelyi, Matthew, 128
Erlacher, Daniel, 137
Ersner-Hershfield, Hal, 68
ESP Game, 216–219
Excitation, 110–114
Exercise, 17, 66, 87, 152, 166–167

F
Faces, 13, 26, 100–104, 151, 184–185, 214, 227
Fading, 265–266
Fairy tales, 143
False memories, 38–45
Familiarity, 31, 40–41, 94–95, 247
Fantasy. *See also* Make-believe
compulsive fantasy, 157–163, 182
dreams and, 143, 157–164
imaginary companions, 91, 196–212, 246
imagination and, 19, 70, 186–187, 224, 254, 267
memory and, 44
pretense box and, 48–52
reality and, 206–210
sexual fantasy, 254
Farver, JoAnn, 211
Faw, Bill, 18
FBI, 115
Fear memory extinction, 148
Fiction, 46–55, 89, 97, 143, 156–161, 186–187, 199–205, 213–215, 246. *See also* Novels
Finding Nemo, 46
First-person perspective, 66, 77–78, 129, 166
Fisher, Helen, 183
Flanagan, Owen, 140
Flow concept, 88–89
Fluency, 40
Flying dreams, 145, 186–187
Forebrain, 132–134
Foreshortening, 8, 239
Forgiveness, 69
Format, 46, 232
Forward model system, 114–120
Foxwell, John, 199
Francis, Anthony, 49
Franklin, Nancy, 28–30

Free will, 79–80
Freezing response, 144
Freud, Sigmund, 41–42, 183, 230
Freudian psychoanalysis, 41
Fright Night, 51, 54
Frost Moon, 49, 53
Full-world imaginings, 99, 227, 252–253
Functional hallucinations, 98
Future, imagining, 59–76
Future-self, 65–68, 164

G
Gaesser, Brendan, 82–83
Galileo, 97, 195
Galinsky, Adam, 188
Galton, Sir Francis, 244
Garry, Maryanne, 40
Geology, 244
Geometric hallucinations, 104–106, 109
Gerlach, Kathy, 71
Goals, achieving, 70–71, 85, 155, 190
Goals, visualizing, 69–75
God, 107, 117, 198, 204, 209
Gods, 106–107, 131, 198, 203–204, 225
Godzilla, 47
Golf, 86, 167
Gollwitzer, Peter, 70
Goodwyn, Erik, 129, 131
Graceland, 18, 172–173
Gratitude, 69, 86–87
Greeks, 90, 203
Green, Melanie, 88
Grief, 72, 91–93, 119
Grinding teeth, 132
Guardian, 129–130
Guha, Ramanathan, 52
Guided imagery, 39, 42, 106–107, 177–178

H
Hallucinations
  about animals, 100
  auditory hallucinations, 90, 96–97, 109, 248
  autoscopic hallucinations, 98
  belief and, 105–108
  color and, 98–102, 111–113, 120–121
  complex hallucinations, 96–103, 109–110, 250–251
  control of, 111–115
  dreams and, 47, 90, 93, 100, 108, 120
  emotion and, 111
  excitation and, 110–114
  forward model system and, 114–120
  functional hallucinations, 98
  geometric hallucinations, 104–106, 109
  grief and, 91–93
  imagery and, 90–110
  imagination and, 90–121, 227, 230, 246–254
  inhibition and, 110–114, 119
  with insight, 92, 105–108, 247–251
  interpretation of, 97, 106–117
  meaningfulness of, 249–250
  memory and, 44
  migraine headaches and, 103–106, 109
  olfactory hallucinations, 104–105
  patterns and, 133–134
  perceptions of, 198–206, 261
  personal significance of, 249–250
  proprioceptive hallucinations, 94
  reasons for, 108–110
  schizophrenic hallucinations, 97–102, 106, 205–206, 249–252
  synesthesia and, 120–121
  tickle hallucination, 116
  types of, 90–121
  unconscious hallucinations, 253–254
  verbal hallucinations, 97, 113, 117–118, 248–251
  visual hallucinations, 97–98, 100–104, 109, 117–118, 249
Happiness, 60–63, 84–89, 155, 162–163, 179–183
Harry Potter movies, 187
Hartmann, Ernest, 130
Hashish, 98
Heal, 56, 177
Healing, 165, 177–181
Health, 66, 83, 87, 92, 97, 112, 118, 176–181, 209, 260
Heart rate, 166, 183–184
Heautoscopy, 98
Hepburn, Audrey, 26
Hippocampus, 124, 151–152
Hobson, Allan, 146
Home economics, 175–176
Hope, 69
Host, 91, 204
Hull, John, 244
Hunger, 105
Hunter-gatherer, 141, 144
Huttenlocher, Jannelen, 30
Hyperphantasia, 16–19
Hypnogogia, 103–104

Hypothetical situations, 7, 14, 19, 39, 45–54, 65, 82, 146

I

Iconic memory, 266–267
Ideas of reference, 99
Illusions
of animals, 235–236, 264
dreams and, 147, 159
feelings and, 94–95
imagery and, 231–239
of independent agency, 199–200
memory and, 37
optical illusions, 55–57
perceptual illusions, 110–111, 118
Ponzo illusion, 8
Imagery
color and, 13, 236–239, 262–265
conscious imagery, 74–75, 240–243, 247, 253
control of, 102–103, 159–161
daydreams and, 152, 159, 163
description of, 2–7
detail in, 14–19, 48, 65, 101–106, 178, 215, 229, 244–252
dreams and, 126–145
fading of, 265–266
feelings and, 82–87, 182–186
guided imagery, 39, 42, 106–107, 177–178
hallucinations and, 90–110
healing and, 165, 177–181
illusions and, 231–239
improving, 244–245
memory and, 36–45
mental imagery, 2–18, 22–23, 45–46, 84, 90, 101–105, 135, 165–166, 195, 222–223, 229–267
motor imagery, 15–17, 144–145, 166, 177, 239
reading and, 2–3, 46–55, 88–89, 156–161, 215–219, 255, 264–266
reinterpreting, 263–265
rendering, 262–264
sexual imagery, 169, 172, 241–242, 254
spatial mental imagery, 94, 134, 220, 234–243, 252, 262–264
spontaneous imagery, 3, 230, 248
subliminal imagery, 18, 128, 184–185
technology and, 213–228
treatments, 178–179
unconscious imagery, 240–243, 247, 253–254
visual buffers, 22, 262–267
visual mental imagery, 3, 14–22, 36, 81–82, 135, 159–169, 178, 220–223, 232–265
workings of, 229–267
Imaginality, 228
Imaginary companions, 91, 196–212, 246
Imagination
animals using, 267
auditory imagination, 15–17, 105, 194–195, 248
complexity of, 250–251
control of, 92, 201–202, 247–251
daydreaming and, 60, 97, 132–135, 150–164
description of, 1–19, 229–267
differing aspects of, 247
dreaming and, 4, 15–16, 122–149, 186–187, 205, 208, 213, 227, 243, 247, 250–252
explanation of, 1–19, 229–267
fantasies and, 19, 70, 186–187, 224, 254, 267
feelings and, 77–89, 182–186
of future, 59–76
hallucinations and, 90–121, 227, 230, 246–254
as healing, 165, 177–181
home economics and, 175–176
imaginality, 228
imaginary friends, 91, 196–212, 246
improving, 244–245
involuntary imaginings, 16, 148, 242, 247–249
localization of, 247, 252–253
memory and, 20–58, 170–175, 255–258
as mental training, 165–195
mind-wandering and, 71, 150–164
mnemonics and, 138, 168–174
morality and, 77–84
for pain relief, 69, 177–181
perception and, 20–58, 167, 192, 261–267
process of, 254–255
science and, 190–195, 213–228
as self-improvement, 165–195
technology and, 213–228
voluntary imaginings, 92, 103, 247–251, 254
workings of, 1–19, 229–267
Imagination inflation, 40
Impact bias, 72
Implanted memories, 40–41
Improvisation, 148–149, 185–186
Improvisational theater, 133, 148, 186
Index card, 170
Inferior parietal lobe, 134–135
Inferotemporal cortex, 102
Inhibition, 110–119